Tying Down the Wind

Jeremy P. Tarcher ❧ *Putnam*
a member of
Penguin Putnam Inc.
New York

Tying Down the Wind

ADVENTURES
IN THE
WORST WEATHER
ON EARTH

ERIC PINDER

Most Tarcher/Putnam books are available at special discounts for
bulk purchases for sales promotions, premiums, fund-raising, and
educational needs. Special books or book excerpts also can be created
to fit specific needs. For details, write Putnam Special Markets,
375 Hudson Street, New York, NY 10014.

Jeremy P. Tarcher/Putnam
a member of
Penguin Putnam Inc.
375 Hudson Street
New York, NY 10014
www.penguinputnam.com

Library of Congress Cataloging-in-Publication Data

Pinder, Eric, date.
Tying down the wind : adventures in the worst weather on Earth /
Eric Pinder.
p.   cm
ISBN 1-58542-060-3
1. Meteorology.   2. Washington, Mount (N.H.)—Climate.   I. Title.

QC861.2 .P48   2000   00-037368
551.5—dc21

Printed in the United States of America
1   3   5   7   9   10   8   6   4   2
This book is printed on acid-free paper. ∞

BOOK DESIGN BY DEBORAH KERNER

Acknowledgments

I am grateful to all of the meteorologists, technicians, mountain guides, educators, and friends who offered comments and suggestions during the years in which this book was being written.

Sarah Curtis, Dave Thurlow, Mark Ross-Parent, Gloria Hutchings, Lynne Host, Dar Gibson, and Jacob Klee freely shared their insights and anecdotes about mountains, Antarctica, and weather in general. Without their contributions, this book would have been much more difficult to write and much less entertaining to read.

For hospitality, humor, and inspiration, I'm indebted to the entire crew of the Mount Washington Observatory, past and present. For years of support and encouragement, I owe thanks to my parents, Richard and Jane Pinder, as well as to Sarah Shor, Tim Ewald, Steve Piotrow, and Meredith Piotrow. Special thanks also are due to Mitch Horowitz at Tarcher Putnam, Barbara Shor, Susan Ross-Parent, Britt Scharringhausen, Jennifer Morin, and many others who read drafts of the manuscript and provided valuable feedback and constructive criticism. Most of all I wish to thank my agent, Laura Langlie, for her advice and enthusiasm.

Contents

Introduction

MARK TWAIN ALWAYS HAD PLENTY TO SAY ABOUT THE weather, particularly New England weather. "Cold!" he once wrote. "If the thermometer had been an inch longer we'd all have frozen to death."

Perhaps his most famous quotation is a phrase he borrowed from Charles Dudley Warner, a Hartford, Connecticut, newspaper editor, in 1897: "Everybody talks about the weather, but nobody does anything about it."

Twain wasn't the only one who thought that way. An old joke still making the rounds in the North Country (which humorist Bill Bryson recalls in *A Walk in the Woods*) is about how winter seems nine months long, and is followed by "three months of not-so-good sledding."

I've lived in the Northeast all my life, through heat waves and blizzards, thunderstorms and freezing rain, so I'm no stranger to the vagaries of weather. But even as a child I always wanted to know more. In answer to Mr. Sam Clemens' complaint, I didn't want just to talk about the weather—I wanted to do something about it.

"What is wind?" is a question I first asked myself—with the intention of writing about it—while walking through a woody corner of Massachusetts on a frosty autumn evening at the age of 24, listening to invisible breezes sift and surge through the pine branches. This book is the end result of my attempt to find the answer.

IN THE SPRING of 1995, I left Massachusetts and seized an opportunity to investigate the wind like never before. Where better to study the science of meteorology than on a rime-encrusted mountaintop at an observatory devoted specifically to that purpose? So I journeyed to a summit where the wind never stops howling and snow falls in July.

My intention was to stay for only a season (the "not-so-good sledding" one, which most people call summer). Little did I know that for years to come I would live there—and once almost die there.

Something in the wind gripped my imagination and refused to let go.

PART I

An Ocean of Air

Born in the Belly
of the Sun

"IT WON'T BE LONG BEFORE YOU GET HAMMERED BY A hundred-mile-an-hour blast," meteorologist Mark Ross-Parent shouted, cupping his hands to his mouth like a megaphone. His thick, reddish-blond beard bristled in the wind. "My first week here we had a peak gust of one-sixty!"

I barely heard him. A sudden gust of wind made me stagger back and trip over a metal grate. Mark's mouth formed another string of syllables, but they were muffled by the growl of the air. "What?" I hollered back at him. He spoke again. Whatever words he uttered slipped away in the wind long before they reached my ears.

"What?" I roared back a second time.

Mark yelled, almost screamed, and I finally heard a whisper siphoned through the gale. "I said 'watch your step!'" He carefully enunciated each word.

The two of us were standing—or trying to stand—in sustained gale-force winds on top of Mount Washington, New Hampshire. The land around us crested and dipped with blue and brown mountains, fading on the horizon at the limits of vision. It was a cold, blustery May morning in 1995 (or so I remember thinking. A few months of acclimatization would soon alter my perspective). Sixty-one years earlier on that very spot the strongest wind ever measured on Earth—231 mph—shrieked across the summit. Three excited meteorologists had witnessed the event, and had escaped with minor injuries, a deep respect for the destructive power of wind, and a tale they would tell proudly all their lives.

Ever since that day, technicians at the meteorological observatory had continuously documented the extremes of weather. Mark and I were among them; our job was to watch the sky—and sometimes just to watch out!

On that brisk afternoon, we were dressed in coats, wool hats, and gloves. Mark, showing his rebellious side, wore short pants. It was almost summer, after all. He gripped the railing of the watch deck and braced against the powerful impact of quintillions of air molecules against his chest. The hood of his jacket started to snap back and forth in the wind. He has a short, stocky, strong frame—the perfect build for a weather observer at one of the windiest locations in the world. Tall people soon learned either to stay indoors or else resign themselves to the inevitable scrapes and bruises from being batted around by the paw of the wind.

I had just started my new job as a weather observer and already had a sore elbow to show for it; an unexpected gust had slammed me against the metal door. I still had a lot to learn.

Mark was showing me the ropes. Or rather, the lack of ropes. "When the wind's really blowing, they'd just whip around and hurt you. Or you'd get all tangled up." I took his meaning: in a hurricane-force wind, hold on tight and don't let go.

A few bright patches of rime ice decorated the wind-scoured summit cone. The high albedo, or reflectivity, of the white ice crystals bounced some 90 percent of the sun's incoming energy back into space. On the limited surface area of a mountaintop, the effect was minimal—the snow and ice patches were exposed and vulnerable to any sharp increase in temperature. A river of air, thousands of miles long, poured constantly over the peak, but the individual air molecules in that swift stream were scarcely affected by their brief contact with the alpine ice and snow. However, in places and times that the snow cover was more extensive—like the long winter of the last ice age—a different story unfolded. The white, icy surface radiated away longwave energy—heat—exceptionally well. It also reflected shorter wavelengths of light, including ultraviolet (UV) rays, which otherwise would have contributed to warming the ground. In this way, the ice perpetuated itself. The white surface continued to keep the air cool by nullifying the warmth of the sun. (In modern times, the same process helps produce bitterly cold nights across North America each winter, whenever there is no cloud cover to "trap" the radiated heat.) Only a climate shift, or the sudden arrival of a warm air mass from beyond the horizon, could break the cycle.

Cycles and change—that was what weather was all about. I had come here to study weather at its worst in the hopes of understanding weather at its best. My firm belief was that no matter how severe the weather, the underlying rules it follows always stay the same and are, to a degree, predictable. Over the next several years, I planned to put that theory to the test.

Any good science book can tell you how the sun heats Earth more in some places and less in others; in doing so it creates weather. Probe deeper into the science of meteorology and you may learn how the sun makes cotton-ball cumulus clouds materialize in a blue sky— and how it heaves the 20-billion-ton bulk of a hurricane from the surface of a tropical sea. But there are certain things that you cannot learn from a textbook.

I was not satisfied parroting old equations and standard school-room explanations. Ever since childhood, when an unexpected lightning strike blasted apart an oak tree outside my bedroom window and the house shook with thunder, I had felt an urge to understand the hidden forces of weather. I wanted to comprehend fully—to feel in my gut—how the sun was responsible not just for warmth, but also for bitter cold. For example, the icy wind howling at that instant over Mount Washington and past my ears.

In taking this job at the observatory, I hoped to go beyond a mere textbook knowledge of the mathematical intricacies of weather. I needed to acquire some hands-on experience. Literally. Given the absence of safety lines, it quickly appeared I would not be disappointed.

My task at that moment was to take a weather observation for the National Weather Service. Cloud cover, cloud height, visibility, temperature, dew point, and half a dozen other statistics were measured and written down in SA (Surface Airways) code—later changed to METAR* code—every hour, then transmitted via modem to weather service headquarters in Bismark, North Dakota. Thousands of other stations—from sunny California all the way to Antarctica—repeated the procedure, feeding endless data into massive computers in order to determine regional forecasts, thunderstorm potential, short- and long-term climate indications, and even the evidence for a possible global warming trend. The weather station on Mount Washington was just one small but intriguing piece of the bigger picture.

As I walked across the summit cone, stratocumulus clouds billowed below me, threatening to engulf the summit in fog. A glance at the sling psychrometer in my hand told me the temperature had dropped to 35 degrees Fahrenheit. Factoring in the wind speed, the apparent temperature plunged to minus-2. The expression on Mark's face—he was gritting his teeth—told me that he had started to regret his choice of attire. With a wave of his hand—we could no longer hear each other at all over the howl of the wind—he indicated that he would meet me back in the office. I continued to scan the sky.

The sling psychrometer I held was a device used to measure moisture in the atmosphere. It contained a dry-bulb thermometer, which provided the temperature of the air, plus a second thermometer wrapped in a wet wick. The evaporation of moisture off the wick cooled the thermometer, to the point at which no more water could evaporate. The difference between the two instrument readings let me establish the dew point and relative humidity.

Weather is ultimately about the transfer of latent and potential energy via wind and water. (Energy in the atmosphere mostly comes from the sun—but not entirely, as I would soon discover.) The evaporation of water requires an input of energy, or heat, to boost each molecule from the slower liquid phase to a more active gaseous state. It takes 540 calories of energy to evaporate a single gram of water. Whenever a water molecule jumps phase, it "robs" a little energy from

*METAR is a French acronym. Translated, it means Aviation Routine Weather Report.

the surrounding atmosphere. The air cools. The larger the number of molecules evaporating, the greater the drop in temperature. That explains the cool breezes I had so often felt immediately after summer showers—the rainwater was in the process of evaporating even as it fell from the clouds, and on the ground it continued to change into gas. Both evaporation and condensation are always occurring on and above any liquid water—including a raindrop—but the temperature and pressure determine which process is predominant.

On the mountaintop, the wet wick of the psychrometer in my hand operated on the same principle. Water molecules always evaporate faster and in greater quantity in a warm, dry air mass than a cold, moist one. So as molecules disappeared off the wet wick, more and more heat was transferred—"removed"—from the air; the mercury dipped. By determining exactly how far it had fallen and comparing that number to the ambient temperature, I successfully measured the moisture in the sky.

The result did not surprise me. The stratocumulus clouds so close to the summit hinted at the arrival of a humid air mass. Now I had the actual numbers. Relative humidity had climbed to over 85 percent. The dew point—the temperature at which water vapor in an air mass must condense into fog at a given atmospheric pressure—had been rising all afternoon and now stood at 31 degrees, only a few degrees lower than the actual air temperature. Once the two numbers met, the air would be saturated and a blinding mist would appear. So I didn't expect our view of the sky to last very long, and I was not disappointed. Only a few hours later the air thickened into a soupy blur, cutting off the visibility at 100 feet. The sun dimmed and disappeared behind a screen of fog.

That first week on the mountain fascinated me but taught me nothing new about wind and fog. Less than six months later, however, a deadly mountain blizzard would change my perspective forever.

FIRE, AIR, WATER, and earth are the four elements of nature, according to the ancient Greek philosophers. Coincidentally, those few simple words spell out all the vital ingredients in a recipe for weather. A snowflake, a raindrop, a cloud, a tuft of fog, and a soft gust of wind contain a little of all four.

Fire is the element that binds the others together—and, on occasion, rips them apart with a fury unmatched by any lesser force of nature. Since my goal is to explain the hows and whys of weather, I had better start at the beginning—with fire.

More than five years have now passed since I first stood on this mountain and was introduced to the savage wind by Mark Ross-Parent. In that time, I have dodged bullet-sized hailstones in 70-mph gusts, flinched as a hurricane splashed ocean spray against my face, and waded neck-deep through a cloud. Yet I am still awed by the mysterious cycles, loops and swirls that tie the disparate forces of weather into one comprehensive whole.

The sun, of course, is the key.

This afternoon I have scrambled on foot up the mountain to the observatory, arriving on the summit just in time to witness the daily death of the sun. Once each day the furnace of the yellow star effectively switches on and off, alternately heating and cooling half the Earth. That's simple enough. But the method in which it channels weather from place to place takes a lifetime of learning to understand.

As I watch, the life-giving daystar melts and boils on the horizon. The clouds above me are aflame. At this altitude, the sun seems close enough to touch. Like Icarus, who miscalculated the melting point of wax, I start to worry that I have come too close. It is a long way to fall, standing here eye-to-eye with the unblinking sun. But the view!

When I arrive on the peak, a mattress of billowing cumulus clouds floats above the green forest of New Hampshire, kept aloft by a cushion of warm, rising air. Stratocumulus clouds tumble across the sky at an altitude of 4,000 feet—just tall enough for a pair of snow-capped hills to jut like bedposts from the base of the gathering fog.

Overhead, separated from lower clouds by a mile of crisp, clear air, a thin sheet of altostratus flutters in the westerly breeze. Slowly, the higher clouds press down, settling like a blanket atop the cumulus. The sun is squeezed between these two layers. Already low on the horizon, ready to turn in for the night, it sends out a final burst of red rays.

A flickering prairie fire consumes the sky. The clouds broil, a sudden kaleidoscope of red and orange.

Tongues of fire singe the horizon from pole to pole. The sunset becomes a fiery tsunami, flooding the altostratus with orange, and leaving only the black coal of night in its wake. The clouds lunge down and snuff out the flames. In seconds, what little remains of the sunset is doused, and the summit sits snug in the mist of twilight.

TWILIGHT IS A lonely time. The owl, a night dweller, props open her dewy eyes but hesitates to spread her wings and fly. The sunshine is not quite ready to retire; it still flickers awhile below the hills.

As I walk back down from the summit of the mountain, I feel the wind hesitate. It is not quite awake, not ready to stir.

Ghost-like tufts of fog slink across the stones. Their feet sliding silently from rock to rock, these apparitions step lightly on a cushion of air, baffled by my presence here among the clouds. A mountaintop at twilight is no place for the living.

In 1846, philosopher Henry David Thoreau climbed a mountain and stood "deep within the hostile ranks of clouds." He was alone. "It was like sitting in a chimney and waiting for the smoke to blow away," he wrote in his journal.

I sit down to rest on an ice-crusted boulder. Am I the only warm-blooded creature above tree line, a single man in a nest of ghouls? So it seems. The air is empty, blowing cold silence across the boulders.

If I sit still, the ghosts ignore me and continue their march up the mountain. The last ghost in line tugs a ribbon of wind behind him. Sometimes the ribbon slips free from his grip and lashes out, rustling my hair, stirring the cold sedges from their sleep.

"What are you doing here?" the wind whispers. It has spotted me, an unwelcome anomaly in a sleepy world. Then the wind stills, silent, waiting for a reply.

The low sun, filtered through the fog, casts an eerie glow on grasses and stones. It is as if the ground were covered with a fine dust of orange pollen.

Tundra grass, brittle as dry straw, rustles and hisses in harmony with the wind. Autumn is here, and the alpine garden quivers with cold.

I can feel wind tug at my sleeve with unseen hands. When I look for the sun, all that is left is a dying red spray filtering through the trees. For a moment, the fog thins. Stars poke holes through the dark canopy of the evening sky. The air stiffens and chills.

BY CLIMBING THIS mountain I bring myself physically closer to the sun. But a vast chasm still hangs between me and the answers for which I am searching.

All mountains interrupt the relative flatness of Earth's topography; they provoke weather systems to rise and flow in unexpected ways. Orographic lifting forces air to climb and cool adiabatically, abruptly changing the ratio of evaporation to condensation so that a thick fog appears. In simple words, the mountains wring the clouds dry.

Aristotle got this much right. In his often erroneous book *Meteorologica,* from which the science of meteorology gets its name, he states: ". . . mountains and high places act like a thick sponge overhanging the earth and make the water drip through and run together in small quantities in many places."

On distant Mount Rainier, in Washington State, a thousand inches of snow have fallen in a single season, only a short hop— geographically speaking—from a parched, moistureless desert. The dichotomy of a cold, moist climate nestled snugly in the same corner of the map as a hot, dry climate still amazes me. How can that be? Here on Mount Washington, more than five hundred inches of crystalline ice blanketed the peak one winter, while the valley at its base only a few miles away received a tiny fraction of that amount.

This mountain where I monitor the weather is called Agiocochook, "Home of the Great Spirit," in the language of the Abenaki Indians. It is a 400-million-year-old wedge of metamorphic rock thrust abruptly into the sky. "Killer Mountain" is how one book describes this peak. "Home of the World's Worst Weather," boasts another nickname.

Like any mountain, Agiocochook offers weather watchers a stepping stone to the clouds. The summit is a natural laboratory, a platform on which to study the ocean of air that thinly surrounds and protects the globe.

It is also a convenient place to study the sun. The shield of air we call the atmosphere is 18 percent thinner here than at sea level. We are that much closer to the unprotected vacuum of space, to the seemingly empty abyss through which the sun's rays and the solar wind fly unimpeded. Sensitive instruments detect the flux of solar energy as it intersects Earth.

Aristotle correctly asserted in 340 B.C. that there can be no weather without the sun's fire. Of course, he and his contemporaries also got a few things wrong. In one misbegotten book of scientific blunders derived from the dogma of ancient Greece, I remember seeing an unusual theory that made me smile. It claimed that the wind is set in motion by leafy tree branches that bend and sway under their own power, waving the still air like a fan. Why else does the atmosphere swirl and sigh, the ancients pondered, unless propelled by a push from supernaturally animated sticks of wood?

Unfortunately, the unknown author never explained to anyone's satisfaction the mysterious power that makes trees move in the first place. Wiser and better-educated generations laughed at the theory, then discarded it. But Aristotle and his contemporaries still had a point. Even if their answers failed to survive the scientific advances of two dozen centuries, their questions hit the mark today.

Once we know the hows and whys of the gusty movement of air molecules across the sky, once we can explain the clockwise swirl of sinking air around a high-pressure area on a cool, sunny day, a single unanswered question remains: What is it that first stirs the wind— what starts it going?

Wind is the power of our atmosphere in action, the invisible but always driving force behind the beauty of nature. Cold fronts punch through the lower troposphere—the lowest, densest layer of the sky—powered by bitter northwest winds. Warm fronts glide gently over ridges of cooler air, bringing soft rain that lasts hours or days.

But the wind has no earthly origin. Forget the swift push of air that makes a flag snap against its pole. Pay no heed to the breeze that provides lift for the wings of birds, or the howling, moisture-laden gales of the sea that carry pollen and life from one continent to another. Despite the wind's obvious impact on our planet's biosphere, its story is set in motion far, far away.

If we seek to discover the true origin of wind, we must first shrug off the comfortable embrace of Earth's atmosphere. The trek will take us through the vacuum of space, beyond the orbit of the airless moon. We must streak millions of miles past the sulfurous skies of Venus, inward toward the craterous dust of Mercury as it bakes near the sun. Soon, we find ourselves plunging toward the gravity well of a blazing star. We must shield our faces from the heat and willfully dive into gaseous flames far too hot to liquefy. They undulate across the outer shell of this middle-aged star as it drifts silently through space along the cold, dark edge of a spiraling arm of the Milky Way.

Even then, our journey is not over. In a quest to locate the origin of wind, we must sink deeper into the thermonuclear core. For the wind on Earth is born in the belly of the sun.

The sun is an atmosphere so massive that its own weight has set it ablaze. It is a 27-million-degree inferno, hot enough to smash hydrogen molecules together at unimaginable speeds until a new element—helium—emerges from the flames. Nuclear fusion powers the sun. It feeds on itself like an undying coal, consuming the binding energy of atoms for fuel. The sun's kiln is a medieval alchemist's dream.

Our sun is a star, all but immortal in human terms, gobbling up the passing millennia like kindling. Knock tiny Earth loose from a stable orbit and hurl it into this giant furnace and it will disappear like an ice cube dunked in boiling water. The cold Atlantic and the deep bowl of the Pacific will evaporate in half a second. The North Pole's ice cap will shrivel and wink out of existence—a white hat whisked away in a fiery breeze. Rocks and mountains will melt into lava. Continents will turn to slag and dissolve.

And yet, at a safe distance of 93 million miles from Earth, the sun nourishes life. Its caress gently warms the rocks and stirs the breeze. Heat is radiant energy; it jump-starts the motion of air. It supplies the oomph to change water from solid to liquid to gas. Trees sway, animated by the breeze, and not the other way around. Liquid dew evaporates off grass stems and adds moisture to the sky. Cumulus clouds surge upward to cap a pleasant day; high cirrus clouds redden the horizon at dawn and dusk, heralding the weather of tomorrow to the watchful eye.

WIND IS A tool used by nature in a hopeless pursuit of equilibrium. The sun upsets the balance. Except for negligible amounts of heat generated by the molten core in the interior of Earth, by radioactive decay, by the pull and tug of the tide, and by a secondary radiation emitted from the atmosphere, the sun is the source of all heat and energy on Earth. In a single day, sunshine intercepted by our planet provides the equivalent of 700 billion tons of burning coal. And that is only a small fraction of the sun's total energy output.

But there is a problem. The sun does not distribute its warmth evenly. Dark mud absorbs heat far better than does a snow-covered field, which with its high albedo bounces away the sunlight and stays frigid. Already a disparity emerges; the surface of the planet heats and cools in patches. The equator basks in plentiful sunlight while the poles are refrigerated. Air masses expand or contract in different locales. To complicate matters, Earth's swift rotation creates a series of concentric rings at different latitudes, each spinning through the cycle of day and night—warm and cold—at a different speed. Meteorological charts depicting temperature and atmospheric pressure jump up or fall down chaotically depending on time and location. It is up to the wind to smooth out the bumps.

The wind never succeeds. Never.

If a small parcel of air is heated by direct sunlight, the molecules absorb the energy and expand to fill a greater volume. The air becomes lighter and more buoyant; it quickly rises through the cooler, denser air surrounding it, leaving a near-vacuum in its wake. But nature abhors a vacuum. (Aristotle got that right, too.) More air rushes in from all sides to fill the hole. Wind starts to flow.

I can see this process illustrated perfectly on the mountain slope below me. Much of the ground is covered by a lingering snow, but I notice a solitary shelf of smooth, dark rock. The sun bakes down on that shelf, which heats the air above it like a stovetop burner. A thermal is created, a river of warm air rising vertically into the sky.

A second small air mass—a cooler one—slides horizontally off an adjacent snow patch to replace the air that is rising. In time, it too will warm on the rocks and start to rise. But even before that can happen, a third air mass blows onto the vacant snow patch abandoned by the second air mass. And then a fourth parcel of air from a more distant snow patch drifts over to take its place. By heating that one small shelf of rock with its low albedo, the sun has set in motion a chain reaction. Air gusts from place to place, trying to equalize the constantly changing atmospheric pressure. I imagine the repercussions of that thermal over Mount Washington rippling from here all the way to Mount Everest and beyond. The wind never stops, not as long as the sun shines.

AIR IS NOT solid but a gas and therefore breathable, a blessing we owe to the sun. The sun replenishes the sky with heat on a daily basis. Without such a regular downpour of radiant energy, the polar ice caps would grope around the globe with icy paws until at last they touched and clamped together. In the embrace of perpetual winter, life would first retreat to a perilously thin band of warmth at the equator, until crushed at last beneath the weight of glaciers.

Snuff out the sun like a candle and Earth's own inner heat will quickly prove inadequate. Oxygen liquefies at 297 degrees below zero Fahrenheit. Air yields to solid ice at minus-360, a point at which both oxygen and nitrogen crystallize and collapse, tumbling from the sky like snow.* The 400-mile-thick atmosphere quickly compresses, ac-cordion-like, into a thin, deadly frost on Earth's now bitterly cold sur-face.

As the atmosphere dwindles, the wind withers. The oceans freeze solid. First a thin white sheet will appear on the surface; then waves will petrify, stiff and motionless like a snapshot. Ever deeper the ice presses and grows down into the murk where eyeless fish swim in a panic. A few bastions of life survive close to the heat of underwa-ter geothermal vents. But in time, the last drop of liquid water hard-ens into stone. Life in the sea is extinguished.

On the surface, the wind becomes visible for the first time—but no one is left alive to see it. Air condenses to droplets; the atmosphere briefly ripples and flows. Streams of liquid air cascade over dead con-tinents, rippling down ice-plugged river basins. Soon these new streams, too, freeze solid. The airless vacuum of space settles now like a cold blanket on the land. The wind is gone forever.

MUCH LIKE WIND on Earth, the ocean of plasma deep in the sun is fluid-like—it rises, stirs, and flows. Heat from deep in the sun's core surges upward like a geyser through the radiative and convective zones to breach the photosphere, the cooler surface layer we see—but dare not look at directly—from Earth.

Sir Isaac Newton stared for hours at the sun's image in a mirror in 1663, hoping to decipher and understand the nature of light. He tried to drink in the sun's secrets with his eyes. A knife of light and heat carved the sun's image on his retinas. He did not go blind, but salty tears of pain streamed across his pale skin. "To recover the use of my eyes [I] shut myself up in my chamber made dark three days to-gether and used all means to divert my imagination from the Sun," he wrote in warning.

The photosphere—the outer rim of the sun—burns at a maxi-

Nitrogen begins to freeze at a slightly warmer temperature than oxygen, minus-344 degrees Fahrenheit.

mum of only 11,000 degrees Fahrenheit, cooler even than lightning in the stormy skies of Earth—almost mild in comparison to the inferno at the sun's core.

The sun is a yellow orb that never blinks save twice a day, at dawn and dusk. We see only the outermost edge of the daystar. Sometimes the sun's face is blemished by dark pools of sunspots, relatively cold but at 6,700 degrees still furious enough to melt skyscrapers into puddles of steel. The wispy gasses of the corona, visible only during a solar eclipse, burn even brighter, at three million degrees.

Here at the photosphere is where the forces that produce wind and weather launch themselves into space toward the tiny blue and green satellite called Earth, a bright speck nearly lost among the stars.

THE SOLAR WIND is not a wind in the ordinary sense. It explodes as a stream of protons and other charged particles to flood the cosmos. The solar wind radiates in all directions; only a tiny fraction spills out of the sun at an appropriate angle to reach and intercept the Earth.

The solar wind races at speeds of up to 600 miles per second, accelerating as it spews outward into space. A solar flare—a storm of unusual intensity erupting from the sun's depths—increases the power of the solar wind. This sudden outpouring of energy hits Earth like a wave crumbling against the shore.

Solar wind penetrates our upper atmosphere and sprinkles the sky with ions. These charged particles—molecules of nitrogen and oxygen with an unbalanced ratio of protons and electrons—slice through our planet's pale gaseous skin and make the unseen air become visible. They release visible wavelengths of light upon shedding the extra energy and returning to their natural state. We see air molecules hit by bursts of solar energy as the northern lights—aurora borealis—candelabra lit high in the ionosphere, 60 to 80 miles above the surface, where the air is far too thin to breathe.

Alien and full of mystery, the northern lights ignite miles above the highest cloud. These shimmering lights puzzled the ancient Vikings, who believed the aurora originated in mysterious cities at the Pole, earthbound Valhallas inhabited by the gods. Who could have guessed that what they saw in the icy night sky above the arctic circle was actually the earthly residue of monstrous storms on the sun?

Sheets of dancing flame fill each aurora. Viewed from a mountaintop, the aurora shimmers and gleams, a thin veil of incandescence wrapped around the occasional shooting star.

THE SUN'S RAYS hit Earth squarely only at the equator and in the tropics. There, the sun perches in the zenith at noon and easily penetrates the atmosphere with abundant heat. At the poles, the angle of incidence is much greater (due to the curvature of Earth and the 23.5-degree tilt to the planet's axis) and the sun's rays hit the ground at a slant. The sun appears much lower in the sky; as a result, sunlight makes a longer passage through the atmosphere and its energy is diluted. The oblique angle at which it strikes Earth's surface also spreads its energy over a greater surface area, bringing the region little warmth. Hold a flashlight perpendicular to a wall, and its light will be bright and focused over a small circle. That represents sunlight at the equator. Now tilt the flashlight (or the wall). Suddenly, the same amount of light—the same energy—must cover a larger area. It weakens in intensity, just like the slanted sunlight at the poles.

Air climbs away from the warmth of the tropics, launching itself into the sky with a burst of energy, then sinking again near the poles. Cold and warm air masses rub together above the surface, pushing and shoving for territory like armies at war. The ebullition of their contact is what we call wind.

Although the planet's rotation siphons the wind off in odd directions and makes it swirl, in a process we call the Coriolis effect, it is the uneven heating of Earth's surface that sets the whole process in motion. The wind stirs under the yellow eye of its creator, the distant sun.

———————

ONE MORNING I saw a layer of altocumulus press down on the rising sun like a thumb, squashing it into the shape of a square. Pinched between the clouds and the ground, the sun ruptured; red and yellow streams spilled into the sky. At last, singed by fire, the heavy lid of clouds blackened and lifted away, allowing the sun to roll off the horizon. And so the cycle was renewed, and the endless chase of dusk and dawn revived for yet another day.

2

Pinwheel

As DAY YIELDS TO NIGHT, BOTH THE WEATHER AND THE MOOD of the sky must change. An entire hemisphere points to the stars, effectively switching off the sun.

Night is the time of screech owls and shadows, of grunts and stealthy footsteps by unseen creatures. Human beings have always feared the night—we lock our doors, shut our eyes, and hide. But why?

We cannot escape. The planet spins eastward along the equator at 1,038 mph (1,670 kph), a rotation which whisks away the blue hue of daylight and the lingering vermilion sunset, replacing both with an utter absence of color. The entire circumference of 24,900 miles turns once. The air vanishes, gone, invisible. A solid black dome appears in its place, speckled by starlight, lit only by the subtle radiance of the Milky Way.

"Twinkle, Twinkle, Little Star" is a old French folk song readapted by Wolfgang Amadeus Mozart, who wrote new words in German and composed several piano variations before moving on to bigger and better orchestrations. But what causes stars to twinkle in the first place? Mozart probably didn't know or care, and most of the children still singing his nursery rhyme don't stop to wonder. But the question has some bearing on the science of meteorology. It illustrates a curious and important fact about the 400-mile-thick atmosphere twirling endlessly around our bodies, rising far overhead to the brink of space.

The starlight we see takes a long journey before it enters our sky. Waves of light spurt off the photospheres of distant suns and streak across the universe for uncounted millennia. Eventually, inevitably, a few of these tendrils of luminosity enter our solar system: photons from Betelgeuse, Rigel, Polaris, Vega, and the 2,500 other stars visible to the naked eye, plus billions more that require magnification to be seen. Silently, they all pass through the stacked layers of air that surround the Earth: the exosphere, thermosphere, mesosphere, stratosphere, and troposphere.

No one hears starlight arrive and, for most of the sleeping inhabitants on the dark side of Earth, no one sees it. But careful eyes can still detect the Earth's atmosphere after dark. It shimmers. Above it, the stars seem to twinkle and gleam.

The arrival of starlight is announced by visible fluctuations as light waves travel through banded layers of air, each one separated from the others by sharp changes in temperature and density. At each boundary, light refracts at an angle, in much the same way sunlight will bend if it passes from air into water. Place a spoon in a glass of water, and it will appear disjointed at the liquid's surface—an illusion. In the sky, many such boundaries exist. Far above us, the loops and meanderings of a thousand coils of wind crisscross the sky with increasing complexity. And somehow the glimmering starlight must cut through. These feeble beams, which have shone steadily for a trillion trillion miles, in the last ten miles of the journey suddenly wink and quiver in the night, bent and refracted by changes in the temperature—and therefore in the thickness, the density—of air. In space, stars do not twinkle. The fact that they tremble so suddenly in the night sky tells us that our atmosphere is intact. It tints the heavens with little color and no warmth.

I watch for a while as the flickering beacon of Polaris hangs dim and cold in the north. It winks sleepily. The temperature tonight is chilly for July. Cold Canadian air sinks through the troposphere, spreads south, and flows down the slopes of mountains along the east coast of the United States. It pools in sleepy valleys, including the tree-rimmed village in New Hampshire where I live. Sluggish breezes stir unseen, unfelt, in the hour before dawn.

"RED IN THE morning, sailors take warning" is an old adage that glides easily off the tongue, as well it should; generations ago, the ability to read the sky's mood in colors and hues meant life or death. A fisherman at sea who ignored the warning signs of a looming storm often lost more than just his catch.

"Red at night, shepherd's delight" is another old saw. But why is a red sky at night sometimes a boon, a hallmark of good weather, while the same crimson clouds at dawn bring danger?

Even the Bible brings up the matter. In the King James version, in Matthew 16:2–3, no lesser authority than Jesus remarks, "When it is evening, ye say, It will be fair weather: for the sky is red. And in the morning, It will be foul weather to-day: for the sky is red and lowering . . ."

This morning at the eastern edge of the forest, the sun's rays cut through the atmosphere at a slant. They illuminate wispy clouds called cirrus uncinus in the west. These ice-crystal clouds ripple and flow four to six miles above the surface like airborne tresses of gossamer hair, stroked and combed by the jet stream. Beneath them, but still far above the ground, altocumulus clouds advance from the western horizon. Air molecules and microscopic water droplets scatter the low, incoming sunbeams enough to accentuate one extreme of the spectrum, painting the clouds red. Combine that observation with the fact that weather systems in mid-latitudes usually travel west to east, nudged along by the planet's rotation and the direction of the jet stream, and the mystery is solved. The angled rays of the easterly sunrise strike and color the advancing edge of a storm system approaching from the west; it has yet to arrive.

At sunset, the opposite holds true. If the sun sets in the west and illuminates high clouds in the east—or else it pinkens a dry, dusty air mass in the west carrying no threat of rain—chances are that the storm has already passed us by. The moisture-laden clouds continue to push to the east, out to sea, and a less-humid air mass takes their place.

Sailors and shepherds aren't the only people concerned with colorful hues at dawn and dusk. "Evening red and morning gray, two sure signs of one fine day," reads a popular rhyme from Europe. "A red sun has water in its eye," I remember hearing as a child. "Evening gray and morning red sends the traveler wet to bed. Evening red and morning gray sends the traveler on his way." Even William Shakespeare monitored changes in the weather. In *Venus and Adonis*, he writes:

> *A red morn, that ever yet betokened*
> *Wreck to the seaman, tempest to the field,*
> *Sorrow to shepherds, woe to the birds,*
> *Gust and foul flaws to herdmen and to herds.*

This morning, my earnest intention to climb a mountain may be stymied by the very color of the sky. In Shakespearean terms, the red dawn I see bodes ill. A low-pressure system is approaching from the Great Lakes, so the clouds will gradually thicken and lower until at last they dump moisture in the form of rain on the umbrellas of the Northeast.

In preparation for this deluge, I stuff a bulky waterproof rain parka into an easily accessible pocket of my backpack. I would prefer to hike light, but better safe than sorry. I head out the door, walking-stick in hand. Today's events will test the accuracy of old proverbs about red skies. Will I need my rain gear, or not? Folklore and the wisdom of past generations say yes.

As I leave, a friend provides yet another example of folk wisdom, arguing simply, "If you take it, you won't need it. But if you don't, you'll wish you had it."

So perhaps I'll try an experiment: weather folklore versus Murphy's Law. Which will prevail?

AT 5 A.M., when the cock crows and songbirds chirp and chatter, their melodies reverberate in waves. Twilight thaws the sky. Sunshine extinguishes the stars, one by one. Soon the first gush of blue soaks the air. The clouds blush.

My itinerary today is to leave Crawford Notch and hike across the backbone of the Presidential Range, taking pictures as I go—up Mount Eisenhower, Franklin, and Monroe, and at last to the Lakes of the Clouds, perched above tree line in the shadow of Mount Washington. The sharply angled rays of the sun at this early hour offer the best opportunities for photography. As I watch, the sun pours down with a steady supply of sharp white flame. But I don't see pure white—white is really all possible colors mixed and jostled together—nor do I witness a splintered rainbow of hues. I see a simple yellow orb in a blue sky.

Air is the substance of wind; it inspires a daily cycle of color from black to red to blue. Throughout the day, the wave frequency of blue light causes it to bounce and ricochet off oxygen and nitrogen molecules more often than any other color. "Preferential scattering" is the technical term. Ping! Bounce! Backward and forward, right and left—and soon the whole sky is frenzied with blue. The only way to escape is to duck your head and shut your eyes. So dominant is the blue wavelength that for most of each day the supposedly whitish sun burns with a soft, steady yellow, deprived of its azure hues.

"Why is the sky blue?" children ask their parents, who often stumble and stutter over an answer they learned decades ago and promptly forgot. The process is called Rayleigh scattering. Blue light has a shorter wavelength than red light, which causes it to scatter more easily in the atmosphere. The simplest way to picture this phenomenon is to employ an analogy. The smaller a person is, the bigger small objects appear. Just imagine a tiny ladybug crawling across a wool carpet. (I could see one on my living room floor as I packed my gear this morning.) To a tall human being, the rug looks flat and smooth; it is easy to walk across. But the ladybug discovers that the rug is a landscape rippling with hills and valleys—each carpet fiber is a mountain to traverse. The ladybug is small enough to detect every thread.

In the atmosphere, blue light flows with a wavelength of 0.4 micrometers, among the shortest in the visible spectrum. It is therefore short enough to "bounce" off the tiny oxygen and nitrogen molecules. Because the wavelength of red light is "larger" at 0.75 micrometers, it cuts through the clear air unimpeded—like a human striding across a rug.

A more interesting question is: Why is the night sky black? I was tossing and turning all last evening, somehow unable to sleep because I knew I must wake up early to prepare for this hike across the Presidential Range. To pass the time, I looked up at the stars and wondered.

An astronomer named Heinrich Olbers first asked the question, now known as Olbers' paradox, in 1826. To him, a dark night sky seemed impossible. If the universe is all there is, forever unchanging, and the stars continue to pump heat into the cosmos for all eternity, then there should be no way to "lose" excess heat. The universe does not come equipped with a heat sink. Therefore, the nighttime sky should be blazing with light, as more and more heat pours out of the stars with nowhere to go.

But the sky *is* dark at night. Why? Olbers speculated that interstellar space is full of dust which absorbs the energy of distant stars. That provided only part of the solution. The true answer did not arrive until the middle of the twentieth century, with the Big Bang theory and our awareness of an expanding universe.

The universe "cools" itself by occupying a larger and larger volume. Meteorologically speaking, the explanation to Olbers' Paradox is related to the question "Why does air get colder the higher up it goes?" In Earth's troposphere, air cools adiabatically by expanding as it rises. Simply put, as a parcel of air reaches higher elevations, the number of molecules surrounding it decreases; as a result, so does the pressure. Elbow room appears, and the molecules spread apart. But expansion is motion, and motion requires work. Work is energy, and energy is heat. By expanding, the parcel of air cools naturally—and internally, with no loss or gain of heat to the surrounding air. It is a simple dictum of physics, vital to understanding weather on Earth. The dry adiabatic lapse rate of air causes it to cool 5.4 degrees Fahrenheit for every 1,000-foot increase in elevation. "Moist," saturated air cools at a rate of 3.5 degrees per 1,000 feet. And on a much grander scale, a similar principle affects the heat output of stars in an expanding universe. A static universe—one which never changes size—should eventually blaze with light and heat in all directions, day or night. I prefer the universe we have.*

So far this morning, the blue tint of daybreak is pale and weak; it has not yet fully spread across the sky. The hour is still too early. In the east, deeper shades of red, orange, and yellow converge around the upper crescent of the newly risen sun. Cloud droplets, too small to see easily with the naked eye but still larger than air molecules, help scatter red wavelengths. Overhead, the sky's rim now presses down on the sun and pinches it between two hills. The orb bulges, like a fruit squeezed for its pulp. As the sky clamps down, the colors of dawn spray in all directions.

CUMULUS CLOUDS STAMPEDE across the skyline a few hours after sunrise, but so far my rain gear stays safely stowed in my backpack. Red and orange hues long ago faded from the sky. The air is now a smooth, watery blue.

I pass a hand through a parcel of air in front of me and try to imagine the ping-pong cacophony of blue wavelengths swirling through my fingers. The action creates a cool breeze, evaporating a layer of sweat clinging tightly to my skin.

I have climbed halfway up a mountain in a corner of New Hampshire, hoping for a better view of the sunset, which is scheduled to arrive punctually at half past eight. My camera and tripod are dutifully at the ready. But the sun still has a long length of sky to tumble and roll through before it lands again on the horizon.

---

*The solution to Olbers' paradox must also take into account the universe's finite age. We can only see to a distance of about 12 billion light-years; therefore, we are not observing the light of an infinite number of stars.

The trail winds up to the summit, a light-brown strip of color distinct from the surrounding gray rock, eroded by the footsteps of thousands of hikers who have trodden the path before me. I scan the white banners of clouds for signs of the advancing storm system. Cirrus uncinus clouds quickly yield to an icy gauze of cirrostratus far overhead. Wind whispers out of the southeast, never a good sign. In fact, a halo of light now circles the sun, refracted by six-sided ice crystals in the clouds. The optical phenomenon is a harbinger of a warm front, signifying a warm air mass slowly overtaking and sliding over the top of cold air. It is sure to bring precipitation. "The circle of the sun wets the shepherd," I recall.

High on the mountain, hovering at eye level, a harmless cluster of cotton-ball cumulus belies the threat of higher clouds and the insinuations of the southeasterly breeze. According to the National Weather Service forecast, a storm will hit the mountain not long after the sun falls asleep at half past eight. Hopefully, that will allow me and dozens of other hikers plenty of time to get to shelter—if reports over the weather radio, interrupted by loud bursts of static, stay true. But nightfall will turn fierce and rainy with all the heat and fury of July. Much to the disappointment of shepherds, a red sky at dusk is unlikely to occur.

As the hours pass, low-level cumulus clouds cast shadows in the valleys and darken the lower hills. We hikers have climbed among them. Just as young children stand on tiptoe to touch a leaf on a high branch, so we now stretch our limbs and try to brush against the woolly bellies of clouds.

Standing at an altitude where the cumulus clouds frisk and play, I shield my eyes from the sun and peer into the east. Clear air stretches for 95 miles. But below me, a milky gray undercast—stratocumulus—fills the entire valley. Rippling clouds separate distant peaks like islands, drowning the lowest hills in fog.

"Doesn't it look like we could just walk across the clouds from mountain to mountain?" sighs a tall hiker with a red bandanna wrapped around her forehead. As she passes by, she points with the tip of her walkingstick to the nearest peak, thrust like a volcanic island above a choppy sea of cloud.

IN SCIENCE, THE answer to any single question is likely to evoke at least two more—it is hopeless to try to keep up. As soon as puzzling rhymes about red skies and sailors are answered and satisfied, new inquiries emerge. For instance, why do storms move from west to east in temperate latitudes? Why not east to west, or north to south? How do weather forecasters know if a storm developing over Lake Superior will move east and trouble us here in New England, rather than striking out for some other point on the globe?

Today, the clouds dutifully slide east across the Great Lakes into New York and New England, and a very familiar pattern emerges on weather maps. The Ferrel cell in which we live—between 30 degrees and 60 degrees north latitude, encompassing most of the United States—serves up the prevailing westerlies. Although the sun heats the planet and sets the whole atmosphere in motion, it is the easterly rotation of Earth and the tilt of its axis that combine to tweak the wind and the seasons in predictable ways.

As wind nudges weather from horizon to horizon, it is constantly pulled, tugged, and twisted in several directions at once by forces invisible to the eye. At the equator, the sun's rays bake the surface with hard, flat rays, heating the air until it is less dense than cooler levels of air above it. Buoyant, the warm vapors rise through the troposphere to a "roof" called the tropopause at 40,000 to 50,000 feet. The risen air then flows north or south, cools, and sinks back down. Finally, it rushes across the ground back toward the equator, only to rise once more in an endless cycle.

The opposite holds true at the geographic poles, where dense, sinking air inhibits the formation of clouds and paints a semipermanent **H**, or high-pressure area, on maps of the North Pole and much of the frozen Antarctic desert.

That, at least, is the simple explanation. The truth is more complicated. One mechanically minded meteorologist taught me to think of the atmosphere as a wind factory, an engine of air with transparent pipes and knobs and gears. Cumulus clouds rise vertically and then shrink and dissipate across the globe, like pistons powered by up-and-down surges of heat and air. Sunlight and the condensation of water vapor provide the fuel to keep this engine going. All the while, the planet spins in a slow pirouette around the sun, so that heat is deposited first here, then there. The system is active, alive—never static.

Wind always rushes from high pressure to low pressure, like a rock dropped from a high peak into a low ravine; you expect it to land with a thud. On weather maps, then, wind ought to flow in a beeline, straight from a big blue **H** stamped over Minnesota to a red capital **L** in Ohio. That sounds good, simple and logical—but it's wrong. Many twists and turns complicate the dispersal of weather. For instance, there is a "force" of nature named after French engineer Gaspard Coriolis that always gets in the way.

The Coriolis effect is just one of several hidden impulses that determine the direction of the horizontal motion of air we call wind. It is, perhaps, the simplest force to describe and the hardest to truly understand. Listen to Dave Thurlow, meteorologist and host of the popular radio program *The Weather Notebook,* explain the Coriolis effect in a single sentence: "The Earth spins, so wind spins." In the same breath, he grumbles that any reply so brief barely scratches the surface. We must dig deeper to find a true explanation.

The Coriolis effect or "force" is not really a force, but rather an effect created by the Earth's rotation. Weather books often oversimplify the matter, stating that the Coriolis effect *causes* wind to veer from a straight path, when in fact it's more accurate to say that the apparent curvature of the wind *is* the Coriolis effect.

The Coriolis effect directly influences the plunge of air from high to low. It turns a straight line into an active swirl. Wind bends around low pressure counterclockwise in the northern hemisphere, but clockwise around high-pressure systems, where the air sinks and the sun shines. The faster the wind flows, the stronger the Coriolis effect becomes. It channels cold breezes from nor'easters and the howls of Pacific typhoons; it directs the cloudy arms of hurricanes as they traverse the Caribbean.

At the observatory on top of Mount Washington, I once watched Dave Thurlow demonstrate the effect in meteorological pantomime. He is a tall, dark-haired, bearded man with an obvious enthusiasm for science. Using his body and a conference table as props, he acted out the role of wind traveling from high to low pressure. One small step took him to the edge of the table. He explained, "Pressure-gradient force makes the wind want to go straight from high to low." He jerked suddenly to the right and took another step or two. "But the Coriolis effect bends it to the right." For 20 seconds he continued in this manner: a step forward, then a turn to the right. Another step forward, another turn to the right. He quickly circumnavigated the table, completing a loop in much the same way that wind whirls around the center of a storm. "Pressure-gradient force and the Coriolis effect play a perpetual tug-of-war with the wind," he concluded.* "Pressure-gradient force wins—but only by a little."

That explains *how* Earth's rotation bends the wind. The trickier question is *why*? Most weather books use an old analogy—trying to draw a straight line on a turntable. But since vinyl LP records are almost obsolete, I decided to search for a better metaphor. The answer I was looking for was provided by a young meteorologist named Jacob Klee, who once spent a year atop Mount Washington (where his tall frame and habit of talking too fast earned him the nickname "Jetstream") before moving on to forecasting jobs in Oklahoma and Virginia.

Klee refers to "the near-futility of playing catch on a merry-go-round." It is, as far as I know, a unique way to visualize the Coriolis effect. The edge of the merry-go-round represents Earth's equator, and the center represents the poles. "Someone standing on the edge is rotating faster than the person near the center," explains Klee. If you toss a ball from the edge to your friend in the center, the ball keeps the

---

Friction is the third participant in this meteorological tug-of-war. Friction at ground level slows the wind down and weakens the Coriolis effect slightly.

"twist"—the angular momentum—that it had when it left your hand. "So it will head off, if going clockwise, to the right of the catcher." Likewise, if the person in the center tosses a ball to you at the edge, then the ball has nearly no "twist." As a result, it will appear to curve to your right, and you'll find yourself frustrated trying to catch it. "Neat, huh?" says Klee.

EARLY EACH MORNING, lazy air drifts through the knobby branches of oak, maple, and pine. The sun's rays must reach deep into the woods to pry long shadows off the soil. The shadows shorten as the morning advances toward noon. Slowly the rising sun hauls them in like ropes, heavy and dark. Hours later, as daylight wanes in the afternoon, newer and darker shadows extend from the pillars of trees and lengthen in dark stripes across the forest floor.

Air cools as night approaches. Fog materializes. In simple terms, the air temperature lowers to the dew point, and water vapor changes from a gaseous to a liquid form. (More technically, as energy is removed from the system, the rate of condensation exceeds that of evaporation. I mention this as an aside, because certain weather professionals are increasingly frustrated with the idea of air having a "holding capacity" for water vapor, which is the popular explanation given by many teachers and in textbooks. Cold air is often described as holding less moisture than warm air. The truth is a bit more complicated. "It's like being right for the wrong reason," Jacob Klee once complained.

"But if you were explaining it quickly to a general audience? In layman's terms?" I asked.

"I'd say . . ." He paused. "I'd probably say cold air has less of a holding capacity than warm air," he conceded. So it's a useful metaphor, if unfortunately imprecise. More water vapor does tend to accompany a warm air mass than a cold one, however.)

As the air cools, moisture condenses. In doing so, it releases energy to the sky. A solitary wisp of fog—cumulus fractus—washes up over the mountain on which I am climbing, and tumbles and spins across the boulders. It masks my eyes like a sheet of moist spider's silk.

Dusk is when the sun switches off and the cold stars reignite. It is a time of transition, and an appropriate moment for the storm to strike. A speckle of drizzle dots my arms and face. I am—literally—running late. After arriving a few hours earlier at the unheated alpine hut where I would spend the night, I left again to pursue more photographic subjects farther up the trail. I intended to return earlier, but a side excursion to view the sunset has put me behind schedule. Shelter is not far away. The quick pulse of the breeze accelerates as wind pumps a sudden excitement into the air. The sky wheezes and hisses. My backpack is wrenched off one shoulder by the wind. I pause to tighten the straps.

All mountains jut into the sky and interfere with the normal flow of wind across the globe. A large enough peak can create its own weather—or magnify the power of a slow-moving storm, wringing moisture from the clouds.

Already, the mountain's edge is masked in fog. But my destination lies straight ahead. The dim outline of an Appalachian Mountain Club hut is visible in the distance, wrapped in gray rags of fog. The wood-and-rock structure materializes once and then vanishes like Brigadoon.

I walk faster, wading through mist. So thick is the fog, I'm tempted to try to swim. The water mats my hair flat, until the wind wrings it dry. I see fat beads of liquid flung into the breeze off my forehead; my scalp is a faucet dripping horizontally.

Wind presses and prods my skin, stretches it taut with cold across my cheekbones. The pressure of each gust feels like the misty palm of a ghost who haunts the clouds, curiously probing the outline of my skull.

Condensed water vapor cascades across the mountain, carried along like flotsam in a noisy river of oxygen and nitrogen. I speculate that water vapor fills perhaps two to four percent of the total mass—the wettest the air can get at this temperature and still allow me to breathe. Is it possible to drown in the clouds?

Waves of mist ripple and churn at my feet. When I look east, turning my head into the breeze, the wind tugs at my ear, pulling me along the trail like a tardy child. I wrench my head back to one side, feel my ear settle back against my skull. A tsunami-shaped cloud splashes high and snuffs out the sun; twilight returns. That quickly, I am blind. At last I stumble into the doorway of the hut, glad to find shelter.

My rain gear waits in my pack, unused, proving Murphy's Law correct—just barely. I do not complain. The tangy scent of dinner on the stove wafts through the interior of the hut. Warm, fresh bread sits on the counter. Outside, the first torrent of rain finally splatters against the windowpane. Wind drives it horizontally through an open crack on the bottom, and a cold puddle appears on the floor.

I withdraw deeper into the building. A warmer breeze circles from room to room, mirroring the flow of air in the cyclone outdoors. How long this storm may last, and how intense the deluge of precipitation will become, I can't say. But when in doubt, it's always best to negotiate with Mother Nature from inside four strong walls.

Centuries ago, that wasn't an option—not here. The hazards of wind-blasted mountains were so extreme that Native Americans in New England refused to travel above timberline; it was taboo to set foot on ridges where the trees couldn't grow. How times have changed. Sightseers and hikers now come and go in increasing numbers, and people actually live and work on the summit at the meteorological observatory, documenting the severity of the alpine climate with numbers and codes.

Just two hundred years ago, this mountaintop was barren, hostile, uninhabited, and the word "meteorology" did not yet exist.

3

A Ladder to the Sky

THE MOUNTAIN I AM STANDING UPON INTERRUPTS THE HOR-
izontal flow of air and forces weather systems to swerve and rise—
sometimes in unexpected ways. "It's so difficult to forecast for the
summit," laments one meteorologist, sighing. "The weather service
may be calling for 'in the fog all day,' but just by waking up that morning and looking and seeing what's out there—just by experience—you
can say, 'you know what, we're not going to be in the fog today.'"

The average pressure at this outpost is 23.66 inches of mercury,
or 801.22 millibars, compared to 29.92 inches at sea level. "With that
much less pressure, it's harder to get things done," quipped technician
Lee Vincent, a large, burly, joking man, in one of two self-published
books about life on the fog-shrouded summit.

This afternoon the peak juts above the clouds, seemingly too tall
to climb. Waves of wind wash up the slopes but tumble back, like a
tide crashing against the shore. Even the sun appears to roll off the
summit helplessly into an ocean of mist far below. One by one, the
clouds beach themselves like whales on a ridge of jagged gray stone.

I wade ankle-deep across the boulders in a shallow mist that
swirls like liquid through my toes. Far below me, cumulus clouds
cruise across the skyline. The wind propels them. Each small cloud
contains several tons of moisture wrapped in a billowing white sack.
The average cloud weighs approximately 25 tons. Whenever the air is
unstable—with cumulus clouds rapidly growing taller than they are

wide—much of that moisture falls as showers. Mercifully, it does not all drop at once.

Larger and taller clouds weigh proportionally more. A towering behemoth like a cumulonimbus, dripping with water and spewing hail, can tip the scales at ten million tons. According to today's forecast, we will not be troubled by thunderstorms or showers, only fog. But who knows for sure?

As I step across the glacial scree, the clouds on the horizon swim closer on the breeze. They look harmless enough. Wind herds them west. Shadows hang from the clouds' bellies and drag through the distant trees like dark, clinging vines.

So FAR THE summit itself has escaped the thickening fog, but that will not last long. A patch of clear air hovers around me like a halo, a buffer between the misty sky and solid land. I scramble up to the topmost parapet of the summit's watchtower, a few feet above the incoming fog layer, to search for a view.

A crowd of eight or nine summertime tourists joins me, peering down at a 60-foot drop to the rocks. Beneath us the slopes tumble downhill—at first gradually but then ever more steeply, ending in a drop of 3,000 feet to a giant bowl carved out of the bedrock by glaciers.

I am paid to climb mountains, to trudge dutifully up the snowy slopes and steep ravines, only to collect a paycheck at the top. "It's the high point of my career," I tell friends, who grimace at the pun. Actually, I am now paid to climb just one mountain in particular, to share the pebbles of knowledge I've picked up regarding the peak's weather, geology, and history. The reward for this task is part spiritual, part cash.

I often wonder what brings me to this mountaintop, which like all mountains seems to teeter at the brink of the world, where wind scours the cold rocks and the air thins perceptibly to a gaseous shell between the ground and the vacuum of space. Why is there such an appeal in extremes? I wonder, also, about the motives of the millions of sightseers who go out of their way each year to witness the lofty views and inclement alpine weather so rarely experienced in the low-lands.

Not everyone who visits the summit knows what to expect. "Is the summit on top?" one tourist asked.

Weather observers, park rangers, and other hosts and guides keep a list of humorous questions from the busy summer season, when more sightseers arrive by car than on foot. Often their inexperience shows: "Is walking down called hiking?"

"Are there any restrooms along the trails?"

"Who cut down all the trees up here?"

One man heard his stomach grumble. So he pointed to the tall observatory tower on top of which the anemometers and other instruments operate. "When does the round restaurant open?" he wanted to know.

Each summer a memorial is placed outside on the summit cone at the exact site where 23-year-old Lizzie Bourne died from exposure in 1855—on a cool, foggy day, much like today—becoming the first woman to perish on the "Killer Mountain." A full-size, weatherbeaten duplicate of the memorial is also displayed in the summit museum, next to an exhibit about weather hazards. Like the original, the museum memorial notes that she "died on this spot in 1855." Predictably, someone asks, "Why did you let the lady die on the floor?"

Of course, certain folk know exactly what they're getting into when it comes to weather. An observatory intern from Plymouth State College once remarked, "I became a meteorologist because it was the only profession I can be wrong at half the time and not get fired."

This afternoon's visitors seem a pretty smart bunch, curious about the weather and awed by the view. A large, red-faced man leans out over the brink and peeks straight down to the ground. His belly sags over the tower's edge. He grips the railing so tightly that his knuckles bulge, a ridge of white bone.

Behind me, a gasp of awe issues from his companions; the fog slides open like a curtain to reveal the distant hills. Off in the west, as the sun sinks behind a toothy ridge of mountains, the sky reddens and cools.

The jumble of human bodies separates and slowly spreads around the parapet. Sightseers laden with cameras jog around the circular platform in search of windows in the churning fog.

"The Appalachian Trail runs down that ridge," I say to no one in particular. I wave an arm at the southern tier of mountains, a spine of ancient, metamorphic rock running north up the planet's backbone. Off to the east, a river pours down the landscape, lit by the setting sun. An undercast of low clouds ripples below us like a vast lake. "Back that way's where you could see the ocean, if we could see anything in this fog." I must lift my voice above the grumbles of the rising wind. No one appears to listen.

Mist thickens while I speak. Somewhere up above us, a moist cloud clamps down on the mountain. We are being swallowed by the sky, eaten alive.

Against my skin, the breeze is a cold caress. The cotton-clad tourists start to hunch and shiver inside their thin summer coats. One dark-haired man dressed in a Washington Redskins sweatshirt crouches against the concrete wall, hiding from the wind. A middle-aged woman rubs her hands to keep them warm.

We will have to retreat inside soon, but not just yet. I walk around the parapet one more time and point north, where every now and then a limb of some distant mountain winks into view, a solid arm of rock entwined in a sheet of fog.

WHAT DRAWS PEOPLE to the tops of mountains? Like eagles in our aeries, we proudly survey land and sky, as if we had something to do with it all. Why else do children lie on their backs in the summer grass and imagine shapes in the clouds?

From early childhood to the fading memories of old age, some mysterious impulse pulls our eyes—and our hearts—to the sky. I have always wondered why.

When asked what possible reason a sane person could have for climbing a mountain, George H. Mallory replied, "Because it's there." But what was the real attraction? The mountain itself? Or the rolling world far beneath it, seen all at once from a high peak, like a jigsaw puzzle of rivers, hills, and streams suddenly assembled before his eyes? (Years later, after a successful ascent of Mount Everest, Sir Edmund Hillary answered the same question—why do it?—and revealed himself to be the worst sort of peak-bagger: "We knocked the bastard off," he boasted.)

A quarter-million people scale my favorite mountain each year, tiny and obscure though it seems. A much stronger itch to climb carries a daring few to the thin, icy air on top of Everest, the roof of the world. But you don't need a mountain to make a ladder to the sky; millions of people ride elevators to the observation deck on the Empire State Building just to watch the sunset magnify the building-block silhouette of Manhattan. The destinations change, but the yearning to climb to the top is always the same.

FOR SOME, THE highest point of land is never quite high enough; it fails to satisfy. We always reach a little higher, try to pull the sky down on our heads like a comfortable hat.

As the sun melts and merges with fog, a thin man with wire-rim glasses suddenly ignores his guide—me—and tries to climb higher. And why not? There are still a few rungs of the ladder left to climb, a few inches closer to the clouds.

The man separates from his group in the parapet and pulls himself up to the very top of the tower. "Don't touch the instruments!" I shout in warning. But the wind, even at a relatively mild 50 mph, dampens my voice. The man gropes for a wind vane and touches it with a gloved hand. Fortunately, a convenient gust lashes out and swats him away. To brace himself, the man clutches a metal ring between a radio antenna and the wind vane.

This tower is the highest point of land for a thousand miles, a concrete bastion holding up the sky. From the top of the tower, all of

America rolls away at the man's feet, spilling toward the horizon. No wonder he chooses not to hear me—the entire Earth is underfoot, and he alone stands on a pyramid of rock at its center. Only the clouds fly higher, and even they appear to bend down in homage to the peak. A few strands of mist drag like kite strings across the boulders, a sign that the summit will soon be fully engulfed in clouds and fog.

The man steps forward, wading into the wind. As he leans on a circular ring in the parapet, his baggy pants lash and coil in the breeze, like jib sails cut free. He bends his back, bracing against the breeze. His narrow legs, disguised under a pair of baggy pants, appear to jerk and spasm as the nylon flaps uncontrollably. With a noise like the rotor of a helicopter, the pants lash and flap.

His friends look up and laugh. They grab their cameras. A blond woman's hair rustles like a flag. "It looks like you're a belly dancer!" she shouts. "Or Elvis!"

The man can't hear her above the roar of the wind, so he starts to climb back down again. Each rung on the ladder is a long, slow step, a reluctant return to Earth. He looks puzzled; his eyebrows stitch together and his brow furrows like a plowed field. "What were you laughing at?" he asks. I hear a tinge of frustration rattle in the bottom of his throat.

His friends respond with more laughter. Shielded at last from the breeze, the man's ballooning pants deflate. He drops off the last rung and slips suddenly from the ethereal world of the clouds to solid ground.

ALREADY THE SUN'S orb lolls on a low ridge of hills, resting a minute before it rolls over to the far side of the globe. The red hue of twilight starts to fade, but the show is not over yet. The sky holds one last surprise.

No one else has noticed the speckles of snow, suddenly prickling the sky. As the sun sinks, the clouds drop their burden; a million snowflakes instantly glide on the wind.

Mountains are cloud factories; makers of wind, rain, and snow.

Wind pushes and grunts against the rock-solid slopes, but never budges them. The mountains stand firm, forcing the wind to climb. Like a weary hiker on a summer day, a climbing cloud sweats rain or snow.

As if to demonstrate the principle, a dime-sized flake parachutes past my ear. The tourists soon notice and gasp; their excited chatter cuts an undercurrent to the breeze. "My god! It's snowing in July!" shouts a deep, raspy voice.

The snow squall dances in air, tracing the pattern of each gust and swirl. Snow is a crystal lattice wrapped around a speck of atmospheric dust, holding on for dear life as it plummets thousands of feet to the ground. In summer, snowstorms may rage above the valleys at times, melting in the warmer breezes long before they can touch land. Only on mountaintops does the ground intrude far enough into the icy world of the upper troposphere, giving us a glimpse of Christmas in midsummer.

I peer across the alpine meadows, where a probing wind stretches across the mountain, surrounded by snowflakes. Each twist of wind folds and tumbles the falling snow. Wind strokes the knolls and shoulders of the mountains with a cold hand.

Millions of snowflakes now swivel and dodge above the cairns, but the wind refuses to let them fall; they fly horizontally through the air.

While the snow lasts, white freckles melt on my chin and drip away. I feel a cold sting as each new flake dabs at my skin. One of the tourists, a young girl with red hair, clutches her father's hand. On her toes, she stretches to catch a snowflake.

She smiles as the delicate taste of winter dissolves on her tongue in July.

4

Fire in the Sky

A CLOUD'S ARM REACHES EAST ABOVE A SNOWCAPPED MOUN-
tain and flexes its muscles in the wind. As it extends, the very end of
the cloud splits apart, trying to snatch the last slice of crescent moon
before it sinks below the horizon.

The last embers of sunset still glimmer in the west. I glance east
again, standing atop a mica-schist boulder on the summit, watching
the triangular shadows of the White Mountains slant across the forest
of New England. Minute by minute, the shadows spread thin and
then thicken into night.

A base of cumulus clouds hangs low, scarcely a hundred feet
above the summit, but dark turrets of mist rise high in the churning
air. The latest forecast from the National Weather Service has promised a thunderstorm tonight, and already clouds are starting to cluster
in the sky.

From my perch on the mountaintop, I can see a lid of low clouds
snuff out the streetlights in distant villages. Miles away, the towns of
Littleton and Bethlehem flicker, then vanish. A black veil of cloud settles over the hills and forest of New Hampshire. Up in the sky, the last
few stars wink once and disappear.

Thunder starts to grumble, far away. I fidget, and glance down
the trail toward safety. What am I doing out here on Mount Washington, on the highest point of land in New England, in the midst of a developing thunderstorm?

The air now surges with electricity, and the hair on my head writhes with static. Wind spins and surges. In the distance, huge updrafts of warm air make the clouds balloon skyward.

Soon, lightning starts to grope and claw at the northern mountains, which are still white with patchy snow despite the onset of summer's heat. A knuckle of electricity raps against a crag just below the summit of Mount Adams, three miles away. The storm spits fire as it lurches closer. Sheets of lightning blush inside the tall black clouds.

With each crackle of lightning, the air booms like a giant tympani drum, rattling the mountain's bones. I feel the first spittle of rain trickle down my cheek.

As many as 2,000 thunderstorms may rage across the globe at any given time. An associate once told me that lightning strikes the ground roughly 100 times per second, worldwide. Extrapolating that figure, more than eight million lightning bolts sizzle through the sky each day.

As if to demonstrate the point, a mile-long fork of lightning spears the col between Mount Clay and Mount Jefferson; five seconds later, the sky cackles and roars. Fortunately, I am not alone here; other reckless souls have joined me on the summit. No one has fled in terror— yet. Perhaps that means we are still safe.

"We ought to be careful if the storm comes any closer," cautions Mark Ross-Parent. He is the leader of a group of five people— photographers, meteorologists, and tourists—who have ventured outside to peek into the belly of a cumulonimbus cloud. I walk over to join the group, closer to the safety of the observatory's tower door.

Ross-Parent leans with two hands on the railing and stares straight ahead at the oncoming storm. His thick blond beard bristles in the wind. As he speaks, a second sizzle of energy scars the night sky, perhaps a mile away.

No one dares to move. A vine of electricity drops into the valley, five thousand feet below us, and scrapes across at the ground. I hear a camera shutter click, a second too late. "Damn, missed it," a deep voice mumbles.

Cooler air continues to swirl across the summit, trembling with each blast of thunder.

"That storm's definitely moving closer," Mark warns us again. A loose tuft in the tip of his hat has started to flop back and forth in the breeze. No one responds.

"We really should head in now," he repeats. All around us, the wind is an exhalation. I keep expecting it to pause, to take in a breath before it suffocates, but the lungs of this storm are immense. The wind exhales with furious patience, clearing the way for the storm to follow. Each gust tries to sweep us off the mountain.

Just below the summit, a bushel of black fog rolls like tumbleweed across the plateau. Soon a swarm of predatory black clouds spins around the summit, confronting us eye-to-eye. The wind snarls. And then a boom of thunder punctures the silence, and the metal railing quakes in my hand.

That sudden eruption of noise is enough warning. Without a word, our group shuffles en masse to the door. Behind us, the door clangs shut as the thunder peals.

"There's always one bolt that hits close and sends people running," says Ross-Parent. Our expedition to the edge of a thunderstorm is over, and we now sit safely in the kitchen of the mountaintop observatory, a research and weather station that has been in operation since 1932. A tray of crackers and Swiss cheese sits in the center of the table under a canopy of groping hands.

Fortunately, no casualties resulted from our sightseeing trip. Lightning is never a gentle or playful creature. It is a whip of electricity, a sting of fire five times as hot as the surface of the sun.

A typical lightning stroke burns at 50,000 degrees Fahrenheit; the sun's outer shell, the photosphere, is only 11,000 degrees. Both temperatures are torrid enough to melt—and even evaporate—steel.

Ancient Greek mythology provides us with pictures of a vengeful Zeus, his beard bristling with electricity as he hurls thunderbolts at his enemies. Even today we imagine that heavy rods of lightning fall all at once from the clouds with a thud. But what our eyes see as lightning is actually just the return stroke, a sudden flood of positive charge flowing up from the ground, feeding into the negatively charged stomach of a storm cloud. In fact, the initial thrust of a lightning bolt is invisible—and deadly. It precedes the return stroke by a bare fraction of a second.

People have been struck by lightning and lived, miraculously, to describe the experience. Steel girders have also been struck, and have bubbled into a pile of slag.

Years ago, an enormous willow tree in my backyard fell prey to lightning's fatal touch. The tree was 200 years old, and twice as tall as my house. In a split second, this electric fire ripped open its trunk, gouging out a wooden slab the size of a small car.

I was asleep, but I woke with a start. Out in our backyard, a bomb had exploded. For a minute, an hour—I don't know how long—the rooftop thumped, clanged and rattled as chips of wood and bark and pulp fell like rain. Our family spent most of the next morning picking up the shrapnel.

There is no better place than Mount Washington to experience the gusting winds and ripe rain of a thunderstorm, since a meteorological observatory with a 30-foot-tall concrete tower perches on its summit.

"Want to watch the lightning from the top of the tower?" a meteorologist once joked.

But was he really joking?

On July 9, 1994, a man climbed a metal ladder up to the tower's "Cold Room" and stuck a fluorescent tube out the window. At the time, the winds outside were gusting to 45 mph, and they bit into the weatherbeaten walls like nails. Rain splattered through cracks in the doorway. In the distance, lightning slanted across the sky.

When you tease a lion with a stick, you should expect it to leap up and bite. Thunderstorms are just as irritable. If you stick a glassy tube into the ribs of a cloud, it will pounce upon the intrusion.

The Cold Room is an unheated circular platform just below the parapet. A wall of two-foot-thick reinforced concrete keeps the room insulated and safe during a lightning storm—except when a foolish meteorologist opens the window and sticks out his arm.

By accident, the man was still gripping the tube's metal end in his bare hand when a distant lightning bolt flashed. "Apparently he got shocked," said fellow meteorologist Mike Courtemanche. "You can't fake those facial expressions!"

Courtemanche later joined in the experiment, though he was careful enough to insulate his hand. For 30 seconds he poked and prodded the electricity-laden clouds.

"Sure enough," he wrote in the weather station's logbook, "the top two inches glowed whenever lightning flashed. And that's how I joined the low-IQ club!"

When the storm had swollen to an uncomfortable size, and ribbons of lightning snapped just a mile away, the two observers fled into the heart of the observatory, unscathed.

A comment in the logbook describes the night's activities: "Superb thunderstorms tonight: low-IQ observers play with fluorescent tubes in the tower. If they aren't plugged in, how come they're glowing?"

LIGHTNING IS NEVER plugged in. Lightning dangles off a cloud's frayed edge like an eel. Ravenous, it nibbles at the ground, chews open sockets in rocks and in the trunks of trees. But where does the juice come from?

When a phosphorescent tube is thrust outside the window it produces a blue-green halo, a brush discharge called St. Elmo's fire. It cackles and hisses, almost as if the electricity is sputtering, trying to cough up enough noise to produce a thunderclap.

On May 6, 1937, a spark of St. Elmo's fire exploded in a pocket of flammable hydrogen gas and destroyed the *Hindenburg,* killing 35 people.

In the era of wooden ships, a brush discharge sometimes lit masts and spars like torches—and yet the wood never burned. Sailors thought the fiery electrical apparition was a beacon, a good omen. They named it after Bishop Elmo of Gaeta, patron saint of fire. And in Shakespeare's play *The Tempest*—based in part on a survivor's account of a shipwreck in Bermuda—the spirit Ariel boasts of his electrical prowess: "I flamed amazement," he crows in Act I. "Sometimes I'd divide and burn in many places; on the topmast, the yards, and bowsprit would I flame distinctly."

In truth, the pale flicker of St. Elmo's fire is just a droplet in an ocean of electricity. The air stirs with ions, a thick breathable soup of static charge. A specter of electricity reaches with a ghostly hand for a ship's mast and hisses a warning. And then it fades away, forgotten.

A lightning bolt never hisses; it roars! There is nothing ghostly or subtle about lightning. Each probing finger of electricity rips a hole in the air; it then cauterizes the wound with heat and fire. In less than a second, lightning heats nearby molecules of air to temperatures of 50,000 degrees Fahrenheit. The molecules balloon outward, encounter colder air, produce shock waves, and immediately clap back together with a furious noise—thunder.

I have seen—and heard—enough lightning for one night. A break in the clouds lets starlight spill down onto the summit. In the distance, what little remains of the thunderstorm lurches toward the village of North Conway. A spark of electricity fizzles along the skyline, briefly igniting the border of New Hampshire and Maine for a dozen miles. Then the sky clears. The worst of the storm is past.

SOMETIMES A THUNDERSTORM'S worst violence comes not from electricity, heavy rain, or hail, but from wind. One night, at my home in the valley, I detected scarcely a whisper of a breeze until suddenly the maple tree beneath my window sprang to life, dancing and swaying. A downdraft hit, a surge of cold air flushed out of the middle of a monster-sized cumulonimbus cloud by falling raindrops. It was the morning of July 5, 1999, and the sky was ablaze with meteorological fireworks. Strong winds killed a man in central New Hampshire by toppling a giant tree onto his car. Farther north, wind-jostled tree branches fell on tents and injured numerous campers in the Great North Woods.

The meteorological phenomenon responsible was called a mesoscale convective complex, a tight band of powerful thunderstorms rarely seen in this part of the United States. Days later, I checked at the observatory for a summit's eye view.

"The most striking thing was how quickly the winds picked up. It was the perfect definition of a squall," Tod Hagan tells me. The soft-spoken, bearded computer technician was on duty at the time and witnessed the whole event. Winds were averaging 40 mph that morning—typical for Mount Washington—when suddenly everything changed. "In the space of one minute it went to seventy," Tod says. The peak gust shot up to 88 mph. Tod points to the Hays Chart, a circular wind gauge, and declares, "There's just this vertical spike."

The temperature, by contrast, plummeted by about eight degrees. "When you look at the thermograph, it looks like the temperature just fell off a cliff," says Tod. "The line had been flat and steady."

The storm itself moved across the continent from the Dakotas to Maine at speeds of up to 60 mph. "I had been watching it on the radar for four hours before it got to us. It was moving very quickly. It just marched across Ontario and upstate New York. It was pointed right at us. By the time it got to New York it was in a bow-like formation," or what textbooks call a bow echo. "It was not there, and then it was there."

Sarah Curtis, a young meteorologist with intelligent green eyes and a ready laugh, confesses both awe and enthusiasm for the thunderstorms that hammer the mountains: "Frontal passages are phenomenal, when there's a line of thunderstorms. There's such a contrast in the temperatures." She thinks back to a similar occurrence one summer ago. "It was averaging maybe thirty or forty mph, and when the front moved through it gusted to a hundred and twenty. And within fifteen minutes it was down again. The chart from that is amazing."

An ordinary cold front swept over the mountain that day, accompanied by a run-of-the-mill thunderstorm and a typical downdraft. But that was enough. "Lightning storms up here are incredible," Sarah declares. She nods across the room to her colleague and friend, observer and photographer Lynne Host, a forty-something woman with a glint in her eye and a reputation for mischievous humor. On Lynne's darkroom wall hang two dramatic photographs of lightning strikes close to the summit of Mount Washington—pictures she must have been outside to take. "Lynne runs around on the deck, and I choose not to do that," adds Sarah, laughing. "I stand in the doorway. I tend to be very sheepish when it comes to taking chances."

That's still close enough to enjoy the show. Sarah vividly recalls hearing St. Elmo's fire—the dangerous, high-pitched buzzing noise before the strike—in approximately the same location where Mike Courtemanche saw it four years earlier. "Saint Elmo's fire, when you're out near the deck in the middle of a lightning storm, and you hear the crackling from the air charging up . . ." She pauses in mid-sentence. "It's caused by tiny discharges from the ionized air. That's where the strike is attracted to. That's where the lightning ends up. I haven't seen it, just heard it." It's usually a good indication that it's time to get inside.

The power of thunderstorms was what attracted Sarah to meteorology in the first place. "I was always interested in the weather. As a child, I would watch thunderstorms go through with my dad." She smiles at the thought. "Whenever there was a thunderstorm coming, my father and I would stand at the front screen door, and first we'd feel the spray from the rain and watch the lightning. That was a great memory for me." It happened in her hometown of Merrimack, Massachusetts.

"And then when I was in high school, a friend of mine was teaching me how to surf. So to be able to forecast whether there were going to be good waves or not we had to watch the weather patterns. I had never really paid attention to that, to the details of the weather patterns—how they moved and how they affected the ocean and such."

"You can surf in Massachusetts?" I ask, interrupting.

Sarah laughs again. "Yeah, you can. Well, some people can. I can't. I tried it a couple of times, but didn't end up staying with the surfing." What she did end up staying with was the weather, majoring in meteorology at the University of Massachusetts at Lowell. It was a decision that would, years later, lead her on a career path directly to Mount Washington and the weather observatory on its summit.

IT WAS HERE on the mountaintop that I met Mark and Kim Chagnon, owners of the sailboat *KMS Beagle,* which harbors off the New England coastline. The Chagnons know far more about thunderstorms from firsthand experience than I do; lightning struck their boat not once but twice in the first five years they owned it. They blame this bad luck, half-seriously, on an old superstition.

"Everyone told us, whatever you do, don't change the name," says Kim, a thin, blond woman in her thirties. Disliking the old name, they ignored that advice—and paid the price.

"We figured after we were struck once, that was it. No such luck," says her husband, Mark.

KMS stands for "Kim, Mark, and Stubs," the latter being a seven-year-old terrier, a dog like Toto from *The Wizard of Oz.* The allusion to Darwin's *Voyage of the Beagle* is intentional. "We figured when we sold the house and bought the boat it would be our voyage of discovery. We felt the name was appropriate," Kim explains.

Their first close encounter with the electrical wrath of a cumulonimbus cloud occurred in a marina in Quincy, Massachusetts, in 1994. While Kim slept in the boat, Mark walked down the dock, returning from a dock party where approximately 200 people had gathered. He had almost reached the *KMS Beagle* when lightning hit the mast. It blew off the antennas.

"The people on the next boat—they were standing on the threshold of their boat. They were knocked inside," remembers Mark. Shock waves emanated from the pulse of lightning and gave everyone nearby—including Mark—a rude shove. "I was knocked to my knees. I don't know if it was just fear or instinct."

"We had just put brand-new electronics in—all fried," adds Kim. But then she holds up a hand, correcting herself. "The TV worked horrible before that, but it's worked great ever since. It's the only thing that worked better. I slept through the whole thing—don't remember it at all. But the whole marina saw it."

The next strike provided her with a front-row view. "The second strike was worse," she says with a visible shudder. A storm had just steamrolled through Portland, Maine, bringing thunder and lightning. The city's power went out, and to make matters worse, Hurricane Bonnie was also approaching.

It was August 1998, and Bonnie had already wreaked havoc in North Carolina and Virginia, after grazing the Virgin Islands and Puerto Rico and narrowly dodging the Bahamas. Its future path was still unknown. Forecasters feared that Bonnie or its remnants would move up the Atlantic Coast and hit New England. Little did Mark and Kim realize that the hurricane was the least of their worries.

"We were in Linekin Bay, right next to Booth Bay, which was not protected at all. So everyone told us you better get into a hurricane basin," Kim explains. The basins are areas of calm, sheltered water, rated on a scale of 1 to 5. A rating of 1 indicates little or no protection; a 5 is safest. "The Basin was a five, so we said 'great, we'll go there.' It looked like a lake, almost landlocked." It is also known as Sheepscot Bay.

They were not alone. "When we pulled into the Basin, there were quite a few people in there for the same reason—waiting out this hurricane," says Mark.

At around two or three P.M., all the boats massed in the Basin encountered a rain squall with lightning. Kim and Mark had already invited friends from another boat, the *Katinka,* over to the *KMS Beagle* for dinner. They had anchored the second boat about one hundred yards away. Tall trees—taller than the masts—surrounded both boats at the edge of the Basin.

At the first sign of lightning, Mark put away the propane tanks. He paused to listen to the NOAA weather radio, which announced gusts of 70 mph howling through Portland. And then it struck.

"You almost knew it happened before it happened," says Kim.

Mark nods in agreement. "I remember it being really quiet—then *boom!*" Their guests, Dave and Barbara, concurred, saying they felt it before it even hit.

"Every hair on your body, you could feel it standing on end," Kim tells me. "And then it hit. It sounded like the atomic bomb—a direct hit." Even their dog, Stubs, puffed up like an angry cat. "Then there was a 'pop' and it looked like the Fourth of July. We saw a blue light coming down the stays." That blue light was St. Elmo's fire.

Mark continues the story. "The stainless steel antenna on the VHF, it blew it apart. Lit it up red-hot until it melted and blew apart and rained down on the bimini top and the deck. It put holes in the bimini top."

Burning fragments of the antenna scattered across the deck and throughout the boat. "It came into the cockpit and landed on towels and blankets, started a fire," adds Kim. She shakes her head in amazement. For the second time in five years, all electronics on the boat were put out of action.

A boat struck by lightning twice in five years is extremely rare, according to the Chagnons. "Our friends won't come over to our boat anymore anytime there's a storm," Kim says, laughing. "They say it's us."

Mark comments, "After the first time I never was really scared of lightning, but after you see it firsthand you really . . . appreciate it." He pauses, mulling over the right words. "I'm scared of it now. Never was before."

"It was very nerve-racking leaving there," adds Kim. "We had no electronics. We didn't even have a radio. We could have been killed."

They hurried home the next day, anxious to escape the still-approaching hurricane but now lacking navigational aides of any kind. Waiting was not an option. They knew if the next nightfall caught them still on the water, they would be just as vulnerable to the vagaries of weather as the sailors of past centuries. The only lights available to guide them would be the distant stars.

5

Music of the Spheres

STARLIGHT IS ALWAYS SILENT.

No dialogue exists between red Aldebaran, the supergiant Antares, and Earth. Even a locally important star like Polaris sends no messages; it pinpoints the North Pole and weakly brightens the sky, but adds nothing to the murmurs and whispers of the evening breeze. If the stars speak to us at all, they talk in words we cannot hear. We strain our ears across the void and come up empty.

The distance separating one sun from another is incomprehensibly vast. Even a trek to the nearest star*—a short, interstellar hop to Proxima Centauri and back again—would be long enough to make the combined journeys of all the greatest explorers in human history—Marco Polo, Magellan, Armstrong—seem like casual backyard strolls by comparison. Whether we like it or not, we live in a seemingly empty universe. An abyss. Solid matter is as rare and precious in this cosmos as platinum is on Earth. We live on a "pale blue dot," in the words of the late astronomer Carl Sagan. Our planet is a tiny oasis of life in a sea of nothingness.

Many, perhaps most, people find themselves rebelling against that lonely scenario. It offends our sensibilities. Stare too long into the eternity just beyond the edge of our galaxy, and agoraphobia takes hold.

---

*Other than our own sun, of course.

Are you comfortable with the endlessly expanding, silent universe that astronomers describe? Or do you prefer instead to fill the darkness with radiance, the way a thoughtful painter colors an empty canvas? The way a diarist's pen scrawls wistfully across an empty white page?

Our ancestors chose the latter path. They decorated the night sky with stories and legends, always in an effort to pull the distant starlight somehow closer, to make the cosmos less frightening and more familiar. They conjured up patterns and pictures in the random sparks of distant suns.

Today we know with scientific precision that the tiny flicker of light called Betelgeuse in the constellation Orion is in fact a reddish M2 supergiant, 800 times wider than our own sun, burning itself up at an explosive rate nearly four quadrillion miles away. But long ago, it was nothing but a studded jewel in the vest of Orion the Hunter. Sirius, the Dog Star, nipped at his heel, and suddenly the chase was on. Taurus the Bull fled in terror from Orion's spear. The velvety canvas came alive; stories scrolled across the sky. Simple. Comfortable. Familiar.

In past eras, the heavens shrank in size to accommodate the limits of the human imagination. The ancient Greeks stared up at the 2,160-mile-wide moon and saw an object no bigger than a shield. If they climbed a tall enough mountain and stretched out their arms, perhaps they could pull it from the sky and bring it home. Did chieftains and nobles ever wonder how such a prize might look on their walls?

Of course, no one succeeded in pulling down the moon, if they ever truly tried. But the universe of the past was a cozy one. You could reach out a hand and almost touch the wandering planets, the jewel-like stars.

Today our scientific knowledge opens up increasingly wider horizons. Telescopes and spaceships probe deeply into the mysteries of space, and new wonders unfold. We have discovered much that was missing from the comfortable myths of the past. For example, in all the trillions of miles between here and Proxima Centauri, there exists not a single breath of air. We find no clear confirmation of life other than our own.

So the universe is empty and airless after all. The gem-studded tapestry still speaks only to our eyes. Beyond the thin vapor of our atmosphere, the cosmos is quiet and cold.

"NATURE ABHORS A vacuum," according to the old cliché, and surprisingly enough the evidence agrees. The fact that we can hear those words spoken aloud is proof enough.

Sound waves require a substance in which to propagate. We will never hear the triumphant cries of Orion the Hunter or the frightened braying of his prey; the emptiness of space is a barrier no noise can cross. Wind goes mute at the edge of our atmosphere, silenced and suffocated.

Gusts of wind shriek and howl in the troposphere precisely because the wind is made of air—countless billions of molecules always jostling and squirming. At ground level, where the air is thickest, nightfall becomes a time of sighs and shadows, of bending and creaking trees. Moths with grotesque shapes flutter their wings against the windows and smack into streetlights in an instinctive attempt to follow the moon. In cities, horns honk, brakes screech. Wind mutters. No one escapes the noise.

Tonight, as I sit at home in a green valley in northern New England, I hear the wind speak in two voices, whispering to clouds that blanket the stars. I can hear a dialogue between piercing gusts and weighty lulls. The sky inhales with lungs impossibly large, and then exhales all with a sudden scream. Trees on a nearby streetcorner quaver and sway. Their shadows whip across my bedroom walls like black vines, and the whole house shakes.

Outside, I hear a sudden noise like a high-pitched whistle between gap teeth. Deeper gusts—a storm cloud clearing its throat—boom in the hills. The breeze that carries this cacophony of weather to my ears is invisible, but not voiceless. The air is alive, emotive.

Peeking out the window, I see leaves swirling and rustling against the pavement in the street.

HUMAN BEINGS LIVE submerged in an ocean we cannot see. The atmosphere swirls around us; it fumbles through our fingers, combs through our hair, and seeps deep into our lungs, and yet we scarcely notice. The wind whispers news; sound waves tap lightly on the tympani of our inner ears and register in our brains as meaningful words or music.

Wind is nothing more complex than air in constant motion. Molecules press together around our ankles and then twirl away unnoticed. Take a single step across a beach at sea level and you will kick into the sky quintillions of molecules, all of them pressed so tightly together that less than a millionth of an inch separates one unit of air from the next. These molecules impact violently and ricochet a billion times per second at our feet, but we wade through them effortlessly, unaware.

Far overhead, in the thermosphere at the fringe of space, the air thins and spreads to an almost imperceptible degree. Vast spaces now separate each faint gasp. At an altitude of 350 miles, it takes about a minute on average for one molecule of air to encounter and bounce against another. The composition of the atmosphere is homogenous—a fairly consistent ratio of nitrogen, oxygen, and argon, with trace gasses like carbon dioxide, neon, helium, krypton, and xenon—until the near-vacuum of the exosphere, which is primarily populated by stray atoms and ions traveling at incredible velocities. Unimpeded, gathering speed, parcels of air are flung into orbit. Though oxygen is too heavy a molecule to easily escape the pull of gravity, the very lightest elements—helium and hydrogen—encounter few obstacles to slow them down; they accelerate and may shoot into space, never to return. But no one hears them leave.

More than 99.9 percent of the atmosphere's mass exists below the mesopause, only 50 miles high—an altitude somewhat arbitrarily designated as the beginning of "space." The troposphere, the lowest stratum of air, is where we live and breathe. It is a thick ribbon of oxygen, nitrogen, and water vapor compressed to a depth of roughly seven miles, kept in place by gravity. It contains roughly 75 percent of all the air in the sky. Fluid air rises at the equator, heated by the sun, and sinks at the poles. Clouds form, wind howls, storms rage and die. Sound waves undulate back and forth in the thick broth of the troposphere, and constantly we hear the grunt of effort as wind pushes these weather systems forever across the globe.

AT DAWN I walk alone into Gorham, a tourist town at the foot of Mount Washington and the Presidential Range. A purplish hue thickens on the horizon. Wings of wind flap against my shoulders, my neck, my back.

A sudden gust strums at the laces on my hiking boots. If I look east, the red crescent of the sun peeks over a tree-topped hill and casts its shadow into the west. I must turn my head to see; the wind tugs at my cheek. The air this morning nips with frost, so I pull a scarf tighter over my chin. It's September, after all. A week ago the humidity brought thunderstorms, but now the North Country seems ready for its first taste of snow. The fabric of the scarf rubs and scratches against my skin, producing a rough noise.

Frost cakes the tips of grass stems that are still green but edging toward brown in anticipation of fall. The grass stiffens as if petrified. Nothing alive moves in the chill. Tree limbs sit frozen and still, while brown oak leaves hang limply on their branches until slapped awake by the lazy breeze, creaking and moaning in protest.

I hear a sudden flutter in the woods, a swoosh of echoes—a crow launches into the sky. Sound waves generated by its flapping wings bounce off branches, reflected back into my ears. Like a shadow come to life, the crow leaps high. It flies across the sun and for a split second darkens the dawn.

In front of me, on a tree that looks empty, a dozen more crows suddenly hoist their wings and take to the air. Their appearance is unexpected, like a cluster of dark buds sprouting into flower.

I listen for a while to the caw and chatter of the crows. They sink back into the trees far ahead of me, so that it takes many minutes to catch up on foot. Still, the wind carries their conversations. I hear their black wings flutter and push against the cold troposphere, slapping waves of wind and sound across the unseen ocean that surrounds us all.

I could, of course, try to sneak up on the crows and observe them furtively. How do they behave when no one is watching? What do they say? But no—I lack the skill to walk so softly. A single snapped twig or the shuffle of my shoes would alert them. In theory, even if I could sprint faster than the speed of sound and arrive at the base of the tree a second or so ahead of any clumsy noises I'd made, the swift compression of air caused by my surge across the ground would produce an earsplitting sonic boom. I doubt the crows would stick around to discover the cause.

Sound is sluggish compared to light, but still quick on the human scale. The speed at which it ripples through the atmosphere varies with temperature and altitude. Hot air causes noise to flow faster, so the throaty caw-caw of a raven in a distant tree might reach my ears a fraction of a second sooner in summer than in winter. In mild temperatures at sea level, sound waves push through the air at 1,126 feet (343 meters) per second. A mostly carbon dioxide atmosphere, such as theorists believe existed eons ago at the birth of planet Earth, would force sound to travel more slowly, at 877 feet (267 meters) per second. Dense water carries sound faster; the songs of whales echo for great distances across the oceans at nearly a mile per second.

As the sun rises and the temperatures warms, I don't really expect to notice the very slight change in the velocity of sound, in the speed at which I hear the wind cry. On this chill morning, the crows and I share a quiet, lonely world. Our only companion is the wind. We share the music it transports from ear to ear.

AUTUMN WIND—SHARP, clear, and cold—can crystallize sound. Scattered birdcalls in the forest ring crisply like bells. Even the dull thud of my boots against the packed mud of September enters my ears magnified, intensified. Sound waves slice through the pure air like sunlight through a perfect glass pane.

The breezes sigh, and old trees wail and groan as gusts agitate their somnolent branches. I feel the air bite at my face; it leaves behind a red flush. The sky is shaded by a hint of precipitation, a hard moistness waiting to fall.

The thin crescent moon hangs just above the trees, so low that its bottom edge drags through the branches. Entangled, the moon snares a branch and is pulled to the ground. As daylight waxes, the crescent moon sinks into the lap of tomorrow and the wind murmurs goodbye.

ALLOW ME TO shake off the cold grip of autumn for a moment and think back to the previous summer. Summer, I believe, is the season when the wind talks loudest. Blunt gusts speak in a manner that leaves no room for misunderstanding. During a thunderstorm, when the sky flushes dangerously, the voice of a cumulonimbus cloud compels children and adults alike to flee from the windows in fear. We listen because we have no choice. Sound waves hammer at the walls and deafen our ears.

When summertime air ripens with heat and humidity, the air will quiver with electricity. Lumpy cumulus clouds rocket skyward, metamorphosing before our eyes from harmless cotton balls to gargantuan, growing towers that flatten against the lid of the tropopause. Cumulus clouds grow into cumulonimbus calvus, looming higher than the world's tallest mountain. The clouds speak loudly—according to the mythology of ancient Greece, their voices represent no lesser authority than Zeus, King of the Gods. Thunder is an explosion of noise that no human larynx can ever hope to match. We listen, motionless and apprehensive, until the sound waves subside. Afterward, our ears ring.

The source of all this tumult is the energy factory deep inside a typical 40,000-foot-tall cumulonimbus cloud. As the storm pounces, a sudden downdraft bends and snaps tree limbs with a crack. Wind whittles at the branches and emits a high-pitched whine. Lightning blazes and thunder rolls.

I remember a July night when the soupy blackness of fog shielded me from the storm. There was no view. Staring through the windows of the observatory on Mount Washington, my colleagues and I saw only infrequent glimmers in the mist as lightning fingered distant peaks. The display disappointed us; faraway echoes of thunder were muffled and slow, barely discernible. We soon turned back to work and ignored the storm—until a bolt smacked the ground only 100 yards away. The impact jolted us out of complacence. The earth shook—Zeus wanted us to pay attention.

Suddenly, it was day; the night fled shrieking. Not even a noontime sun could burn so brightly. Fog vanished suddenly like steam evaporated by a fire. I looked out across an expanse of cold boulders and—so it seemed—newly formed land, etched on the night's canvas.

I half-expected to see the rocks melt into a puddle of lava, so piercing and hot was the light. A white circle obstructed my vision in the center of my eyes; it was an illusion burned into the retina, as if I had stared too long into a lightbulb. Almost immediately, the roar of thunder hammered at the windows. Half a second later, the night clamped down. Darkness and fog oozed in to fill the void cleaved open by lightning.

Dramatic as the lightning was that night, what I remember most is thunder; it shook the marrow of all our bones. We had merely glimpsed lightning with our eyes, but the deadly electrical discharge did not breach the observatory walls. By contrast, thunder injected a barrage of sound waves through the windows, walls, and floors, cutting a swath through our bodies, a swift blow impossible to fend off. Our hearts lurched. An unseen giant had loomed above us and screamed.

Mark Ross-Parent, who had been asleep, wandered upstairs into the weather room only moments after the shattering blow had numbed our ears. He rubbed his eyes sleepily and asked, "What was that noise? It sounded like you were smashing furniture on the floor."

Thunderstorms are the true musicians of weather; percussion is their forte. We listen passively, in awe.

6

Fresh Air

AT SEA LEVEL, DEEP IN THE TROPOSPHERE, A QUANTITY OF air sufficient to fill an ordinary room weighs about two pounds. That's a kilogram, more or less. And that's assuming, of course, that you can catch all the dancing molecules in a net, clump them together, and plop them on a scale.

Earth's atmosphere, taken as a whole, weighs more than five quadrillion tons, or 10,000,000,000,000,000,000 pounds. Tumble enough of the imperceptible particles of oxygen, nitrogen, carbon dioxide, and water vapor together, and finally they add up to something real. Chicken Little need not have worried; if the sky ever does fall, we'll know about it.

Swirling, zigzagging air presses heavily against our bodies: 14.7 pounds per square inch is the accepted figure at sea level. The air permeates our lungs, but we seldom worry about how much it weighs or where it has been. More than one meteorology book states with authority the fact that a single molecule of air from Julius Caesar's dying breath is inhaled into the lungs of each and every one of us, at least once a day. Caesar's last sigh, as Brutus struck him with a knife, has spread across the millennia, carried on the breeze. A typical human breath contains countless billions of molecules, so there is plenty of Caesar to go around.

Air is invisible, unless saturated with smoke or fog or haze. Wave a hand in the air and it obediently parts before you. You see nothing, but the air is far from empty. What else, besides the breath of murdered tyrants, floats in that sea we call the sky? What secrets does the wind carry?

The air in New Hampshire this morning is dense and dry, too cold to carry more than a trace of water vapor, but still aswirl with all the elements necessary to sustain life. Each crisp breath I inhale contains millions of billions of tiny molecules and atoms, all occupying just a few liters of space. For every trillion or so of these molecules, the typical human breath includes perhaps 780 billion molecules of nitrogen, plus another 210 billion bonded pairs of oxygen atoms. They all seep into my lungs, saturate my bloodstream, and swiftly course through my body. Without this constant resupply of air and oxygen, I could not so much as lift a finger, or activate my brain long enough to be even dimly aware of the last stages of a slow suffocation.

As I walk along the outskirts of town, the wind tweaks playfully at my face, dispelling my fears. I sip the sky. Inhalation and exhalation require no conscious thought—instinct monitors my respiration for me. The cool, crisp oxygen floods down my windpipe into my lungs; it divides and spreads quickly through the tubes and channels of my circulatory system. It disperses throughout my torso, arms, and legs, finally merging with mitochondria in every cell. I feel a burst of energy. When a chill breeze suddenly siphons through the pine trees, I rub my hands together for warmth and step more briskly across the ground. (Or do I? For a chilling moment I see an image of myself as a mere puppet wrapped in skin—propelled by the wind, which reaches in like a puppeteer's hand to animate my limbs.)

All living land-based creatures on Earth drink the atmosphere in great gulps. Specks of pollen, dust, and bacteria jumble together along with trace amounts of neon and helium, argon, and krypton, and some carbon dioxide to fill each precious gasp of life. Fully 78 percent of the air we breathe is harmless nitrogen, but the truly important ingredient is oxygen—volatile, eager to combine with other elements, giver of life.

THE MOON, BY contrast, is a dead world; no gasses or moisture of any kind whisper across the parched lunar ravines.

Folklore instructs us that a crescent moon tipped toward the ground will spill rain, like water from a bucket. In this case, folklore is wrong. No precipitation of any kind pours earthward from the moon, nor does Earth's satellite trigger storms, though it does influence the tides. A heavenly slice of green cheese seems a curious place to look for rain, in any case.*

The moon shines down on us, a splotchy, bleak echo of what Earth might have been. Stripped of clouds, water, soil, and the rippling green grass that hides old stones, would Earth look any different? What went wrong on the moon? Or perhaps we should ask instead, what went right on Earth?

As the morning hours expire, I watch the waning moon drag through the upper branches of trees like a slow scythe. Its crescent blade slices at the far horizon. With a silent swoop through maple and birch branches, it sinks below the canopy—out of sight.

Our planet and the moon orbit each other closely as they dance and spin in gravitational lockstep around the sun. On average, they rest only 238,866 miles (384,400 km) apart—a negligible distance in astronomical terms. So a comparison of their skies is worthwhile.

The moon's weak gravity long ago lost its grip on the lunar sky. Whatever thin gaseous shell the moon may once have possessed has long since leaked into space. The solar wind now drills into the soil, unobstructed by clouds or air or magnetic fields; many millions of protons and ions bombard each square inch of the moon's surface every second. The moonscape is dramatically punctured by meteor impacts large and small, wounds that heal ever so slowly as the eons unfold. No wind, rain, or weather erosion exists to smooth the scars.

If you peer up at the cratered surface of Earth's only natural satellite, chances are that it looks exactly the same as it did the night before. Phases may wax and wane, but the stony-faced visage of the

---

*The changing lunar phases do correlate slightly with terrestrial weather, but the reasons for this correlation are unclear. Call it a meteorological mystery.

moon stays rigid and sterile, a timeless sculpture in the heavens. Moon dust catches the sun's rays and bounces them back toward Earth like a 2,160-mile-wide celestial mirror.

The moon has no atmosphere, and therefore no weather.* If meteorologists ever make it to the moon, I suspect that whatever other hazards they face, the greatest danger will be pure boredom. The job will be too easy. I can picture the first lunar forecaster staring sheepishly at the gray lunar dust, announcing the weather in a monotone: "Clear skies, not a cloud anywhere, high temperature reaching 273 degrees Fahrenheit, but bitterly cold in the shade. Watch out for those UV rays. For tonight—for the next two weeks, actually—overnight lows will drop into the minus-240-degree range. Winds are expected to be calm for the next five billion years." Except for an occasional meteor shower or slight moonquake, nothing interrupts the tedium. What else is there to say?

No wind murmurs across the dry lunar plains. No rivers dig canyons in the soil. No plumes of snow shoot off the mountain ridges. No ocean waves crumble against the smooth surface of ancient basaltic lava flows, those dark splotches so prominent on the Earth-facing side of the moon; they were misnamed "maria," the Latin word for seas, by Galileo when he first squinted through his primitive telescope in 1609. No breezes stir the dust of the regolith; no grasses or weeds poke through the dirt. Neil Armstrong's bootprints still linger, perfectly preserved, stamped practically forever in the crisp dust. The treads of Apollo lunar rovers will scar the moon's surface for millions of years.

The moon, it seems, provides nothing but a long list of no's. Fling a kite into the lunar sky, and it catches no wind. It arcs slowly outward toward the stars. But at last, as if yanked by a sudden, reluctant pull

---

*No atmosphere worth speaking of. Outgassing from the moon's interior contributes to an extremely tenuous and shallow "atmosphere" close to the lunar surface. Energy from sunlight and the solar wind may also "loosen" atoms from the ground and release them into the lunar sky. But as far as weather is concerned, the moon's atmosphere is effectively a vacuum.

on the kite string, the toy is dragged back to the surface by the moon's weak gravity, one-sixth as strong as Earth's. No wonder the atmosphere was shed into outer space eons ago.

Stars sparkle and glimmer overhead even in the sunshine. They flicker just out of reach. Sunsets are unremarkable and slow in coming. The ground is naked, exposed. All of space, empty and grim, wraps around the moon with a cold hug.

Imagine for a moment that you stand on the moon's surface at Mare Tranquillitatis, the Sea of Tranquillity. It is high noon, and the sun appears to hang without motion in the sky. You are unprotected by any spacesuit, exposed without roof or walls. No atmosphere shields you from the influx of solar radiation, and that is what's wrong. Quickly, the sunlight burns your eyes. As you blink and tear, the first drop of moisture in four billion years dribbles off your cheek, but it boils away in the vacuum before it can fall into the arid lunar soil. Ultraviolet rays, unfiltered by air and ozone, cut effortlessly through to the surface and penetrate your body at the speed of light. In such a hostile environment, you would quickly acquire skin cancer—but, if it's any comfort, the complete lack of air means you would not live long enough to suffer or even diagnose the symptoms. Instead, the skin on the side of your body facing the sun bubbles into crisp carbon. Your back, shaded in the near-vacuum of space, turns cold. Worst of all, your lungs rasp and plead for oxygen, but they get no answer. Consciousness dims almost instantly. Your body stumbles and falls against the dust of the moon.

If, by some miracle, you manage to stay alert for just a minute more, you can only lie on your back and stare with glazed eyes at the blue and white globe of Earth, distant and motionless in the black sky. Down there, water gurgles into faraway seas, clouds spin and whirl, rain plummets through the air. Though you cannot hear, soft wind whistles and laughs ever so far away. But soon your eyes close for the last time; the mirage dissolves. Your last sensation is an acute pang of homesickness, a painful nostalgia like no one has felt before. And then you feel nothing at all, forever. The moon continues its slow roll from day to night, undisturbed, emotionless, silent.

BACK ON EARTH, I breathe deeply as the afternoon progresses toward evening. With sudden energy I quicken my stride. The brisk air fills my face with a faint flush.

The waning moon has already set, but it will rise again punctually in 12 hours and 50 minutes, looking much the same as before. Only the phases change, and even they are simple and predictable: full, gibbous, quarter, crescent, new. Why is Earth so different, so alive?

Even during this short stroll through the trees, I can watch clouds overhead race and swirl in infinite patterns. Each new hour adorns the sky with colors and shadows that did not exist during the hour before. Wind tweaks the clouds and artfully molds their shapes as if they were clay.

Pictures of our tiny blue and white planet, taken by geostationary satellites 22,500 miles high, show the power of our atmosphere in action—pushing spiral hurricanes across the oceans, sparkling with columnar blue jets and red sprites off the tops of giant thunderstorms. Far below the veneer of cloud cover, rain tumbles and wind howls.

The troposphere is held snug against continents and oceans by gravity, and this allows it to collect water vapor and dust particles, the major ingredients in clouds and storms. Simple evaporation from the ocean pumps billions of gallons of water vapor into the troposphere each day. The clouds I now see twirling in the wind represent a small fraction of that water. In time, they will rain or snow on the mountains, gush into streams, and surge rippling and whitecapped back to the sea.

The atmosphere that supports life also depends on life to sustain it—a convenient symbiosis painfully assembled over the course of four thousand million years. As I exhale, carbon dioxide whistles through my lips and dissipates, a deadly poison to me and most other members of the animal kingdom. My body wants no part of it. But certain algae and green plants eagerly drink it in. They require the stuff. Their chloroplasts in turn expel nourishing oxygen, which seeps back into my lungs and powers the oval-shaped mitochondrian energy factories in my bloodstream and body.

Less than one-tenth of one percent of today's atmosphere consists of carbon dioxide, and for that all human beings may be thankful. Eons ago, little else filled the sky. No complex animal life existed at the beginning. Nor could any such creatures have breathed if they did exist.

A clear view of the deep past is always hidden from us, except in thought experiments and imagination. How did life change the air? Is it changing still? Suppose we could truly turn back the clock, letting evolution career into the past, back to a simpler era, a lifeless world. Backward runs time, through ice ages and volcanic convulsions emitting gases from deep inside Earth, past mountainous upheavals and the slow, swimming strokes of continents as they paddle across the sea.

Time jogs on its heels across long ages ruled by mammals and reptiles. So far on our backward journey, the atmosphere has scarcely changed. We watch as the ancestors of birds tumble to the ground, slithering into the shadows on stunted legs. Time races back to see the first amphibious creature hobble onto shore and sip a painful breath of air. But time does not stop here—it plucks fossils of trilobites from stiff Cambrian rock, softens them and fills them with life, then sends the animals scurrying across warm, shallow seas. It endures long years of boredom as primitive single-celled organisms duplicate themselves in endless repetition.

Zoom back to the beginning. Set foot on primeval Earth. The time is now 4.5 billion years in the unknown past. Once you magically transport yourself in time, the first sight you see is a planetoid of impure iron, newly solidified and accreted from the gaseous nebula that long ago surrounded the sun. The wind is alien, the landscape stark. Feel free to look about, but hurry—your first suffocating breath may also be your last.

Carbon dioxide and carbon monoxide saturate the sky at this early date. An unfamiliar atmosphere presses on the newly formed landmasses of naked stone. Not a trace of oxygen exists, and even nitrogen is rare. No pollens or spores float on the breeze, for life is unknown. It is quite impossible to breathe, but at least we discover that no allergies can afflict our hypothetical time travelers. Wind siphons the carbon dioxide, hydrogen, and dust in swift gusts that nobody hears.

As millions of years pass, meteorites and comets puncture the atmosphere, injecting small quantities of water-ice and other rare materials onto Earth's surface and into the skies. (A very small percentage of water does indeed arrive from outer space—perhaps the adage about rain pouring from the crescent moon is not totally wrong after all.) Volcanic eruptions spew gases from the hot interior of the planet's mantle, a process called "outgassing." Ever so slowly, the atmosphere starts to change.

Many eons must pass before increasingly complex photosynthetic organisms complete the task of producing an oxygen-rich atmosphere, slowly displacing the effluvium of carbon dioxide. Approximately midway through Earth's history, a protective shield of ozone is created in the stratosphere, an absorbent cushion that spares the surface from the full, fatal intensity of sunlight.

For an incomprehensibly long stretch of time, life continues to change and grow. During an epoch two billion years in the past, a noticeable quantity of oxygen finally begins to accumulate. Only now do the ingredients of the atmosphere begin to add up to a more palatable, somewhat familiar whole.

Four hundred million years in the past, vascular plant life colonizes and dominates the land. Animals later struggle and squirm onto shore. The photosynthetic plant world pumps ever more oxygen into the sky, and the animals eagerly drink it in. By now, our time traveler finds the air breathable. The composition is nearly the same as in modern times. Though dinosaurs have not yet roamed the earth, pressing their clawed footprints in the Cretaceous mud, much of the evolution of the atmosphere has already occurred. Life has fully taken hold; the symbiosis between plants and animals thrives.

Few changes to the atmosphere occur during the last 300 million years. At times it is warmer, at other times, colder. Wind and ocean currents shift and slither across the globe, but the overall composition of the atmosphere is little different. Ice ages wax and wane. Volcanic eruptions and asteroid impacts spew ash and dust into the stratosphere to darken the sky; the air clears slowly.

At long last, the journey is done; the long crawl of years from the distant past is summarized, completed. But the trek described here so briefly was never an easy one. In the beginning, our atmosphere churned and stirred violently under the heated gaze of a young sun. Volcanoes rose from the ground, oceans accumulated in dry basins. Earth, so long ago, was nothing but a bulb of rock adrift in space, a miraculous cosmic seed that would one day spawn life.

Life, in turn, has fueled the atmosphere. The wind I feel this morning in New Hampshire as I walk through the evergreens is the living breath of countless generations since the dawn of time. I breathe the inescapable vapors of the past.

7

Calling the Clouds Names

ON A LONG-AGO JULY AFTERNOON, THE HUMIDITY IN THE AIR pasted my cotton t-shirt to my skin. The sun glared down at me, blistering; I felt like the proverbial ant under the magnifying glass. Even in the shade, air turned wavy and translucent with heat. My vision blurred. As I hurried across the hot gravel in front of the old red barn in the dairylands of upstate New York, a drop of gluey sweat trickled the length of my forehead and dripped down from my nose. I tasted a salty tang on my lower lip.

Specks of hay chaff and dust floated on a barely discernible breeze. Two other farm hands and I were tarred brown with the stuff. It clung to the exposed skin of our faces, leaving only thin white circles at the edges of our eyes. The grass at our feet was withered and browned.

I peered into the sky, cupping my right hand over my eyebrows like a visor. I wanted to see rain clouds, thunderstorms—anything to relieve the fervid intensity of July. Far off to the south, a fair-weather cumulus drifted across the sun. The little cotton-ball cloud looked no bigger than the barn against which I leaned—a dwarf as clouds go—but it was perfectly placed. The yellow orb winked and vanished. Heat lessened; a nearby thermometer dipped a single, merciful degree. Air molecules, momentarily cut off from their supply of energy, slowed and gelled. Then the wind whisked away the cloud to unmask the lidless eye of the sun once more.

I detected scarcely any wind from where I stood on the sunbaked ground; the grass did not even ripple. Nonetheless, the cumulus cloud slid into the east, swiftly propelled by stronger gusts at some unspecified higher level of the atmosphere. It lunged beyond the roof of the barn. And that's when the illusion caught my eye. For an all-too-brief instant, the cloud hung motionless in the sky, a fixed point, while barn and soil careered underneath. I felt the massive Earth gyrate along its axis. My feet wobbled, disoriented. When the cloud finally glided out of view, Earth stilled.

That cloud was just an airborne sac of condensed water vapor, a cool pool of moisture tantalizingly close on a woefully hot day. But it flew out of reach, unattainably high no matter how far I stretched my arms. I glanced up wistfully. With a cotton shirtsleeve I wiped sweat from my brow. "I wish I were in a cloud," I muttered.

One of my baseball-capped cohorts leaned against the barn and nodded. "Yeah. It would be so soft and comfortable."

But no. I was longing for tiny droplets of stinging rain, chilled and refreshing. My co-worker imagined a vast pillow. It was a matter of shape versus substance. Science and poetry failed to connect.

Looking back years later, I wonder if both of us weren't right. What is a cloud, anyway? Is it merely a white globule of mist, a random shape of no importance—something to glance at distractedly if, by chance, you happen to gaze at the sky? Or is it a scientifically scripted message in the liquefied air, an indication of what the weather holds in store?

Years of experience at a mountain weather observatory have changed my point of view. I recently read a good science book that described clouds as "weather stations," beacons of temperature, humidity, and wind direction at different altitudes. The shapes we see in the air often tell us what's coming hours or days in advance.

That cloud on the old farm years before probably contained such a message, but I did not know how to read it. Today, perhaps, I do.

IF CLOUDS ARE truly the semaphores of weather, what do they say? I've learned this much, at least: Cumulus congestus clouds are towering white castles of water and ice that fill the sky with architecture; they indicate instability in the air. Pockets of heat and moisture surge upward, causing the clouds to billow and grow. On humid summer days, this means only one thing: showers, and possibly thunderstorms with hail. Better take cover. Altocumulus castellanus, a higher cloud molded by the wind into tiny turrets of mist at altitudes of 10,000 feet or more, also indicates a latent electrical storm ready to spring into existence.

Folklore tells us all we need to know. "When clouds appear like rocks and towers, the Earth's refreshed with frequent showers."

The process is a simple one. As air surges across the flat, heated terrain of a Nebraskan prairie, a volatile geyser of wind and electricity hurled skyward against the force of gravity. Low pressure prevails. The water vapor in the air rises and cools to its dew point, quickly condensing. A flat specimen of cumulus humilis flourishes on the energy of the atmosphere and stretches into cumulus congestus, a cloud taller from top to bottom than it is wide. Always hungry, such a cloud continues to feed on the heat of the prairie. Soon a cumulonimbus cloud—cumulonimbus incus—bulges above the ground like an exclamation point. Booms of thunder vibrate through the swaying grasses and add some oomph to this sudden punctuation of weather. The thunderhead's anvil top flattens against the cold roof of the tropopause at 40,000 feet; streams of ice crystals flow far downwind.

A hundred miles away, these crystals take on sudden meaning to people who observe them from the ground. As the land slides off the prairies, wispy cirrus clouds twirl like apostrophes into the east. If they continue to thicken and spread across the sky, they wave flags of warning in advance of the coming storm.

IN 1802, A London apothecary named Luke Howard followed the example of naturalist Carolus Linnaeus and named clouds in Latin as if they were alive. The highest clouds he dubbed cirrus, or "lock of hair." Today we follow his example and informally call them "mares' tails," fibrous strands of ice crystals combed across the sky by high-altitude winds.

"Mackerel skies and mares' tails make lofty ships carry low sails," say the folklorists. These high, ice-crystal clouds prognosticate warm fronts, bringing sharp winds and rain.

In Howard's system, whole species of clouds evolved. He divided them into quasi kingdoms and phyla, genera and species. Cirrus floccus is the name of a tangled weave of ice crystals stitched across the sky; it indicates slow winds at high altitudes. Such a cloud may tell pilots to expect little in the way of turbulence. Its opposite is cirrus uncinus, or mares' tails, strips of ice elongated by different wind-speeds at the edges of the raging tunnel of the jet stream.

As our knowledge of meteorology increased, and the species of clouds grew and evolved, a legion of tongue twisters entered the lexicon: cirrus cumulonimbogenitus, cirrus fibratus, cirrus intortus—all high ice-crystal formations more than five miles off the ground.

The phylum "alto" signifies a middle-layer cloud, and boasts a rich trove of names. Altocumulus castellanus (indicative of strongly rising air currents and, for pilots, a bumpy ride), altocumulus undulatus, altocumulus lenticularis. This last example, shaped like a lens or a UFO, successfully resists the tugs and pulls of the wind. Unlike most clouds that flow across the sky, lenticular clouds highlight a visible "wave" in the air, an atmospheric echo of the hills or mountains underneath. Wind flows up the crest and cools, condensing to form a cloud. Then it sinks and evaporates on the other side. In this way the individual molecules of air and water vapor that make up the cloud are constantly changing and replenishing themselves as wind pours into the wave at one edge and exits at the other. But the shape of the lenticular cloud itself hovers in place, no matter how strong the wind, while other species of cloud whisk by and disappear beyond the horizon.

Decades ago, a story about extraterrestrials in *Life* magazine featured a supposed alien spacecraft in New Mexico. It was, in fact, a perfect snapshot of a lens-shaped lenticular hovering over the Rockies.

Clouds are nothing more than billowing sacs of condensed water vapor afloat in the troposphere. But that does not stop human imagination from giving them shapes and life. Children point to cumulus clouds twisting in the wind and shout with glee, "Look, a dog!" It is a proud tradition. If we see pictures and myths in the constellations of the stars, why not in clouds as well?

Even the most stubborn-minded adults, who rarely flex their imaginative muscles to the same degree as do children, will at the very least invent taxonomies; we give clouds names and tediously catalog their many guises and moods.

THE LOFTIEST OF all clouds, cirroform, spread in fragile sheets miles overhead and fly at the level of the jet stream. High altitude winds shear off the tops of thunderheads, forming a tress of ice crystals known as cumulonimbus capillatus. They are striated, translucent like thinly frosted glass, and do not hold back the light of sun or moon.

One night, just before a violent but spectacular lightning display, I saw a corona of light rotate quickly around the moon. Circles of green, gold, and red, radiating from the silver disk, made a prism of rings inside rings—a celestial pinwheel that spun as if powered by wind. I swayed on my feet, suddenly dizzy. The heavens reeled.

Of course, the illusion lasted only while the moon was veiled behind that translucent screen of cloud. It was magnified by a papery shade of airborne ice and water; the face loomed impossibly large. So close, I thought I could reach out a hand and touch a moonrock.

Colorful coronas can form due to the diffraction of light through thin sheets of altocumulus and altostratus. But sometimes a high pileus cloud capping an active thunderhead produces the same effect. Iridescence runs up and down the spectrum from red to indigo, different in both size and hue from a true halo. Repeated viewing has taught meteorologists to watch carefully for any corona that expands; it enlarges as water droplets evaporate and disappear, a sign that fairer skies may follow. But a shrinking corona hints of rain soon to fall.

In the language of the sky, a halo also symbolizes rain or snow yet to come. "The bigger the ring, the nearer the rain," a folklorist might say. If a northeasterly wind flows, a halo brings precipitation in less than a day. Sun or moonlight etches a circle in the sky; its radiance is often refracted by the six-sided ice crystals in cirrostratus clouds on the leading edge of a storm.

Our distant ancestors bequeathed to us a rich trove of weather folklore—"Ring around the sun or moon, rain will come soon," states another proverb—but they never knew exactly why a halo brought rain. They only knew that it did.

"Beware a mackerel sky" is another common expression, still murmured by knowledgeable grandparents whenever the bubbling rows of cirrocumulus rib the heavens, looking like fish scales four miles high. A mackerel sky, too, can warn us of coming storms and warm fronts—if we learn properly to read the sky.

FOR AS LONG as weather has played a role in human affairs—ever since the first nomad planted a seed and settled down to be a farmer—people have struggled to predict Mother Nature's moods. All too often, a botched forecast has been the result.

In the litigious United States and other countries, an overzealous lawyer will at times thrust a subpoena in the face of a surprised TV meteorologist whose forecast went astray. In Israel in 1997, a woman sued a weatherman for predicting sunshine on a rainy day. Allegedly she went outdoors unprepared and caught the flu, misled by his optimistic appraisal of the sky. The plaintiff argued that the incorrect forecast had cost her four days of missed work and $38 in medication. Perhaps the meteorologist in question failed to see the words "rain shower" written in the clouds. Or else he misinterpreted the significance of the humidity, or failed to detect a subtle thread of trouble embedded in swirls of color on that day's satellite images. But he wasn't the first meteorologist to guess wrong, or the last. Can you do better?

The blown forecast is a common phenomenon around the globe, repeated day after day like a well-scripted play. Here is one variation:

"You couldn't forecast a shower in your own bathroom!" hollers a red-faced meteorologist at a television station in Boston. (I learn about this encounter secondhand from a snickering colleague; but it rings true.) The man raises a fist in mock anger and wags it like a scolding finger. His embarrassed cohort, sitting at a table far across the conference room, blushes but says nothing. Behind them, bucketfuls of rain splash and slosh across the windowpanes. The rooftop gutters roar like Niagara Falls, gurgling torrents of cold liquid. Lightning flickers once; the sky grumbles and shakes. "My mostly sunny day just drowned," sighs the unlucky forecaster.

To foresee the weather of tomorrow you must predict in intricate detail the movement—the dances, jerks, and skitters—of billions of air molecules from day to day, county to county, state to state. It isn't easy. Consider this experiment: pluck the white globe of a dandelion flower gone to seed and hold it to your lips. Blow once and watch the seeds scatter. Imagine they are clouds adrift on the wind. Can you guess where they will go?

Now multiply the challenge—or the number of seeds—by an order of magnitude and you have some idea of the challenge of forecasting. Chaos theory suggests that a parcel of air faintly stirred by butterfly wings in, say, Pennsylvania, may eventually trigger rain instead of sun in upstate New York—or in Tokyo, Japan. Ripples of uncertainty spread outward with the breeze. We clutch possessively at any forecast that offers us a "60-percent chance of precipitation" or a "partly to mostly sunny day," but certainty eludes us.

Satellite images and computer models can do only so much to decipher the cryptic alphabet hidden in the swath of clouds that circles our globe. Seen from space, white patches of cloud cover half of Earth's surface at any given time like a tattered curtain; the dark bulges of continents peek up through the holes.

EVEN AT AN observatory famous for its wind and weather, sometimes the best we can do is take an educated guess. Tonight the winds gust to 80 mph as a strong cold front powers through the region. At

Mark Ross-Parent's suggestion, the summit crew starts a "peak-wind office pool."

"The winner gets to cook dinner tomorrow," someone jokes. I scribble down a guess of 121 on a sheet of paper. Other numbers range from 85 to 140. If the winds get any higher than that, I figure we'll all blow away, so no one will have to cook dinner in any case. My estimate is clearly on the high end; I base that number on equal measures of isobars and instinct.

One day later I end up a big loser as the gusts never exceed 99 mph. Behind the frontal passage, the barometer rises sharply as the temperature nosedives; the clouds disappear.

The invention of the barometer by Evangelista Torricelli in 1644 opened up an exciting new era in the burgeoning science of meteorology. Suddenly it was possible to weigh air, to quantify the invisible fluid of the atmosphere that no one previously could see or touch, even though they breathed it all their lives.

Oddly enough, when an ancient Greek scientist named Hero first suggested that air was a substance and had weight, his contemporaries laughed. The idea that atmospheric pressure could be used to predict storms did not fully take root until the barometer arrived in the seventeenth century.

The barometer—usually a column of liquid mercury, although Torricelli also experimented with sea water and honey—provided scientists with a reliable gauge. The mercury column rose in fair weather and dropped off steeply in foul, as if shrinking away from the blow of the storm. All at once, the cryptic language of weather became clearer.

Other attempts to measure atmospheric pressure met with no success, at least from a commercial point of view. In 1850, an English country doctor created what he called a "tempest prognosticator," a barometer-like gadget that employed live leeches. Supposedly, the creatures were sensitive to changes in pressure. As storms approached and pressure plummeted, they instinctively scurried inside their bottles; this movement rang a small bell. Predictably, the invention did

not catch on. People in the Victorian age did not care to get their
weather updates from leeches.

Even with the new mercury barometer in hand, the mechanics of
weather were not yet fully understood. A mix of folklore, common
sense, and glimpses of the changing sky made up each forecast, or "indication." Predictions were hit or miss—usually miss. That changed,
in part, when a bald, spectacled man from Philadelphia flew a kite in
a thunderstorm and made a shocking discovery. His name was Ben
Franklin, and his lengthy résumé listed him as a diplomat, founding
father, editor, socialite, and scientist, among other professions.

Franklin's famous experiment with the key and the discovery of
electricity—enlightened though it was—did not represent his only
contribution to meteorology. In October of 1743 he made an observation that turned the folklore of weather prediction into a true science.

One blustery night, Franklin camped outside his home in Philadelphia to observe a scheduled eclipse of the moon. In this frequent
astronomical occurrence, the silvery satellite, perpetually "falling" in
an orbit 238,900 miles away, appears to be slowly devoured by Earth's
shadow. But the eerie spectacle is only a weak echo of a full solar
eclipse, during which a razor-sharp band of darkness cuts through
the daytime air and reignites the stars at noon. Nonetheless, Ben
Franklin and countless others thought the lunar eclipse an event
worth watching.

Unfortunately, he never did see the eclipse that night. What he
saw instead was more important, though he didn't realize the fact until weeks later. His serendipitous discovery revolutionized meteorology.

On that cool October night in 1743, thick clouds clamped over
Philadelphia, driven by a nor'easter powered by moisture from the sea.
The sky blackened; stars vanished. The moon disappeared, snuffed
out by a thickening overcast long before it ever had a chance to fall
into Earth's shadow.

Franklin felt disappointed at first. But then, weeks later, a letter
arrived from his brother in Boston, hundreds of miles north. His
brother described the eclipse in detail, for he had viewed the celestial
event perfectly, seeing through the clear night air at precisely the same

moment that Ben Franklin had stared up at an overcast sky and shaken his fist. Four hours later, the storm hit Boston and blackened the sky with clouds.

This account puzzled Franklin. Since winds in the storm blew from the northeast to southwest, how, then, could it travel in the wrong direction and strike Boston in the north? Shouldn't the wind have pushed the storm the other way?

Whether or not he shouted "Eureka!" at the time, Franklin's keen insight had hit upon the solution: wind rotates in a counterclockwise circle around the center of low pressure areas. While the winds on the night of the eclipse poured out of the northeast into the southwest, the storm center itself swept steadily northward along the Atlantic coast to his brother's doorstep in Boston. It was an astute observation, and all the more remarkable because he made it without the aid of satellite images, without ever seeing the swirling arms of a low pressure system from a vantage point 22,500 miles high.

His next revolutionary idea (pardon the pun) was to recommend that scientists across the country observe and record their local weather at specific times each day, much as he and his brother had done accidentally on the night of the eclipse. By comparing results, surely people could track the motion of weather and perhaps predict where it would go.

Another founding father of the United States, Thomas Jefferson, grabbed hold of the idea. He and a friend took the first simultaneous observations in 1778. By the mid-nineteenth century, while the aforementioned English doctor fumbled with his meteorological leeches, hundreds of weather observers accumulated data under the auspices of the Army Signal Corps in the United States. In Britain, Admiral Robert FitzRoy started a forecasting service of his own. Slowly, science weeded out the puzzles and the mysteries of the clouds.

WEATHER IS THE appearance of what wasn't there before. On a humid summer day I once heard a colleague remark, "It's so neat to watch these clouds just pop into existence." We stared out the window of the observatory into crystal-clear air and 70-mile visibility. Dozens of cauliflower cumulus clouds sprouted in what only moments before had been a clear blue sky.

Whenever a storm thickens in the sky, swept north from Ben Franklin's old cottage in Philadelphia, the atmosphere shifts its mood and presto—a cloud sprouts in the air like a ripe seed. Sunshine, water vapor, and microscopic dust are the prime ingredients in a recipe for clouds; slosh them together in the tumultuous mixing bowl of the atmosphere, stir them with a ladle of wind, and blue skies turn gray.

A world without water would have no weather to speak of: no clouds, rain, or snow. A parched wind might cough and mutter over a dusty, lifeless plain. That's all. Clouds on Earth only appear when water vapor in the air condenses into microscopic marbles of liquid, practically weightless, afloat on the wind. Each individual cloud droplet is only 12 to 50 microns in diameter, practically too small to see. To make a single bead of rainwater requires millions of cloud droplets, cobbled together by the jostling arm of the breeze.

So little water vapor exists above the troposphere, the lowest layer of our atmosphere, that clouds rarely form in higher layers. The lone exception is the occasional cumulonimbus cloud, stretching like a giant to cool its brow in the lower stratosphere, taller than any mountain.

THE BIRTH OF a cloud is an unexpected gift. You turn around, and there it is—a blossom of mist fills what was seconds ago empty air. A bubble of warmth surges upward, cools to its dew point, and the cloud appears out of nothing.

I once watched a perfect specimen of cumulus fractus materialize on a July morning when the air was humid beyond belief—nearly liquid and starting to boil in the summertime sun. As I walked through the woods on the outskirts of Gorham, sweat poured off my skin and flowed into the air like a tributary into a stream. I couldn't tell where my body stopped and the moist air began.

I glanced up and took a reading of the sky. If I was correct, the atmosphere was so hot and humid that the rounded specimens of cumulus I saw hurtling overhead might grow swiftly into more formidable shapes. I knew that quick, heavy showers of rain—perhaps even lightning and hail—could appear within hours. Formerly harmless clouds would bulge toward the stratosphere and reverberate with thunder.

I quickened my pace and arrived home only minutes before the first raindrop splattered noisily on the windowpane. Almost at once, the wet glass blurred. The rain was so heavy I could not see outside. As I slammed the door and raced from room to room to close the windows, thunder grumbled overhead. The weather, sure enough, had followed a predictable pattern.

Clouds are the sky's own calligraphy; they spell out its intent in crisp letters of water and ice. Brushstrokes of mist scroll meaningfully in the breeze. It is up to us to read them—if we can.

PART II

Seasons

8

Tiptoeing Through Autumn

WHEN OCTOBER BREEZES STRUM THE TREES, I SEE MELODIES of maple leaves high in the air. Notes swirl and fall, spilling down the scale *pianissimo* in sun-yellow, red, and orange.

When the wind settles into silence, a ripple of color still thrums across the ground. Puddles of leaves soak the forest floor; golden aspen surges across the soil.

When the dusk of October looms on the horizon, then the last rays of the sun pierce the forest sideways, slanted and dim, igniting sparks in the dark-gray fog. I watch shadows lengthen across the western edge of the Berkshire Hills. In the distance I hear a breath, a sigh, as leaves skitter from branch to mossy earth. In mid-flight, a red leaf twitches, slapping the evening's last sunbeam toward my eye.

Autumn is fleeter than other seasons, squeezed between lingering Indian summers and the onslaughts of early snow. The air stiffens. Breezes drag across the soil, heavy and palpable, weighted with the promise of ice.

I step slowly across the frozen mud of the Connecticut River Valley in western Massachusetts, 150 miles south and west of windy Mount Washington. I've been curious about that mountain and its infamous weather, though I have never climbed the peak—not yet. The year is 1994. A few promontories that scarcely rise above a thousand feet—hills, really—are all the local landscape offers.

Good enough. I'm beginning to realize that you don't need to climb Mount Washington or challenge Everest or voyage to the icy desert of Antarctica to witness both the beauty and the destructive power of weather. The same forces are at work everywhere. Even here, even now.

The grass crunches under my feet. Wind rubs a shiver down my spine; it pilfers warmth from the fibers of my coat, burrows its way through the coils of my scarf. My breath becomes a cloud of cold gray steel.

All across the ground, fallen leaves ignite like a brushfire, but they emit no warmth. Oak leaves sit like black coals amid the flames of red maple and golden birch. Soon their fires fade, crimson to brown to black, while October snuggles against the sleepy shoulder of November.

WHAT IS WIND? My breath floats like a ghostly balloon in the air as I voice that question. A pool of water vapor floods from my lungs, gushes up the warm pipe of my trachea, and finally spills into cold air, where it condenses into a cloud. The mist from my breath materializes like an empty cartoon bubble glued to my lips, filled with a question: What is wind?

Before I can answer, the silver cloud in front of me drifts away on the breeze.

Wind scissors through my jacket and scarf. Puddles of muddy rain gel into slick plateaus on the ground, so that with each new step I slip and slide.

My friend steps briskly ahead of me into the stiff breeze, her gloves scraping a rhythm against the flaps of her coat. She is bundled like a bear cub, wrapped in layers of wool with a soft white hat atop her head like a cap of snow. She calls it "stick season," this slow disrobing of summer, leaf by leaf, till the bones of tall trees rattle and scrape in the wind.

All around, the evening air trembles, cold and clear; even the first stars shiver. In the east, the dying sun shoots a lazy streak of red through the cirrus clouds, a jet of color soothed to blue by the crisp onset of night. A last ember of sunlight burns behind a wisp of cloud. Halos of golden light gleam at its fringes.

IF YOU GROPE for wind with your hand, it dodges and swivels, so that instead of catching the slippery essence of the atmosphere, you come up instead with a fistful of nothing. Wind whistles and mutters in a soft language full of vowels and sighs; it says little that we understand. The wind is alive but alien; it refuses to answer our questions, but asks many of its own.

We query the wind whenever we breathe. What is it made of? Why is air odorless, tasteless, insubstantial—yet essential to life? Will it last forever? If today brings sunshine but tomorrow a dreary gray, what hidden forces move the weather from day to day?

Wind recognizes no boundaries. A pocket of air floating to the apex of Mount Everest will someday swirl across the equator, or rise as steam off the heat of the Caribbean, or else drift to the poles. During the long journey, it may detour through my backyard in New England and scatter the unraked leaves. I may be inhaling such a parcel of air even now. How far has it traveled?

Wind is a fluid, impossible to see yet forceful enough to pry a beachfront dock off its pier. The shrieking whirlwind of a tornado once embedded a straw in the trunk of a tree, like an arrow from a tight-strung bow. Tornadoes have hoisted telephone poles like javelins and launched them across the sky. And in one popular motion picture (frowned upon as "absurd" by meteorologists), a tornado playfully scooped up a dairy cow and spun the puzzled beast in swift circles above a cornfield—while two scientists watched the spectacle, safely inside the kind of protective windproof bubble that only Hollywood can provide.

Wind swings a powerful hammer when angry—yet it can also weaken to a gentle breeze. In autumn, the air twirls in kaleidoscopes of red and orange maple leaves. Wind lets kites float like buoys on the rippled surface of the sky. Wind exhales a deep breath into the sails of ships; it slaps waves across the seas.

Wind inspires many names: in southern California the hot, dry wind that pours down mountains and channels grassfires toward the houses of celebrities will often appear in newspapers; they call it the Santa Ana.

The familiar nor'easters that torment my home in New England swirl counterclockwise around centers of deep low pressure systems that suck moisture from the cold waters of the Atlantic.

Bitter winds agitate the snows of Siberia and are called *buran*. Chinook winds dump moisture in buckets on the western slopes of the Rockies, then drift away as a cool and dry breeze into the east, where they swallow up snow in great gulps. "I've seen those winds— they call them *foehn* in Europe—just eat up fifteen inches of snow. Whoosh! It's gone in no time at all," says a Colorado native. The desert wind that tosses sand into the eyes of camel-saddled nomads in the Middle East is called a *haboob*. And along the Alaskan coast, breezes named williwaws skitter around the fjords, sculpted and shaped by the topography of the 49th state, swirling around islands that jut from the sea.

This evening, a breeze without a name siphons cold air into the valleys. Fog mingles with the stubble of alfalfa in the fields. An inch-long maple seed twirls in the sky, and then is swept far away by a sudden gust of autumn air. Elsewhere the land is still, numbed with cold.

Nothing moves inside the chill; the air solidifies like glass. On the ground, gray coatings frost the tips of grass stems. The stalks stand stiff, frozen. No mosquitoes buzz, no birds sing, no bees fly from flower to flower. No petals or blossoms sway in the breeze. Rigid tree limbs sit frozen inside long, spidery sleeves of ice.

My friend and I are alone, abandoned. Though we know the earth is just asleep, hibernating, it feels dead. Animal tracks—of bear claws, hooves of deer—press like fossils into the hard dry mud of autumn. Over these cavities stand ghostly silhouettes, outlines of empty air.

Frantically the wind pokes and prods this sleepy land, trying to wake something up. Wind whirls around tree trunks, but cannot stir them. It howls against the rocks, but fails to make them move. It lifts up leaves but finds nothing but pebbles underneath.

Frustrated, the wind rips a dead branch off an oak tree and knocks it to the ground; a dry cracking sound snaps in the air.

At last, a piece of shadow springs to life, cut loose from its nook in the trees. It reels into the sky, a slow dark disk—a crow. Three times the crow cries as it flaps overhead. Later my friend and I hear it caw again, deep inside the trees.

So, there is life in late October after all. We are not alone. Satisfied, the wind stills.

My friend stoops to the ground. "A dragonfly," she says. She lifts the insect by its back. The wings shimmer, stiff and lifeless, four chips of frosted glass. Is it dead, mummified by the cold air?

In the distance, the wind roars back to life, a giant sucking of breath. We listen as a wave of wind rolls closer, nudging tree branches out of its way. Around us the air is still.

The giant wave lumbers toward us, so slow we could run away from it if we chose. Instead we wait. It is a giant cresting wave, hundreds of feet high, rumbling through the trees, unstoppable unless at last it crashes against a hillside.

As it nears us, limp leaves dance in a sudden frenzy; they whip and whirl. Dry twigs scrape against the bark. Stiff birch trees begin to bend and sway.

The wave strikes us; my friend's long brown hair whips over her shoulders like river water split by a rock. Shrubs and the stubble of grasses quiver and shake. A bead of rainwater, shaken loose from the trees, lands on my skin. In the air, floating leaves trace the frantic pattern of the wind.

In my friend's hand, the dragonfly is suddenly filled with life. It arches its back. Its wings blur like a hummingbird's. It almost wrenches free from her hand.

"Oh!" She says, and drops it. But as the wind dies, the dragonfly drops back to the soil. The stuff of life did not take hold.

We both peer down. "No, you're not alive, are you?" my friend asks the insect.

But look—a leg twitches. She picks it up again. Definitely, the leg twitches, wiggles, squirms. The creature is alive, then, but dying in the cold clutch of fall. Perhaps it was dead when she first picked it up, as dead as the brown leaves it lay on. But now it lives, replenished for a time by a gust of wind, by the seep of warmth from her hand.

9

Winter

IN A THERMOMETER NAILED TO THE FAR WALL OF THE OLD farmhouse where I am staying, liquid mercury slumps toward the bulb on the bottom. My fingertips brush against doorways and windowsills as I walk room to room; the wood is cold to the touch. When I rake coals in the wood stove, prodding the dying fire back to life, pale red embers wink at me sleepily, then erupt in sudden flame. Veins of frost blush red on the windows. A distant clock chimes six times.

Snowfall, like starlight, is silent. Who knows what happened overnight? Perhaps three inches fell, or 13 or 30. Anything is possible. I awaken early just to see the damage done by last night's blizzard. Gusts of wind still paw at the windows. The walls creak, mumble, and tremble in the cold air.

Outside, the pine trees stoop and sag, hunched under heavy white shoulders of snow. Flurries and snow showers clutter the air with a white, grainy haze; I can't see so far as to the driveway, much less to the silo across the road. I wonder—is the barn still there? Perhaps last night's blizzard has picked it up and carried it to Kansas, or Oz.

I am alone here, a visitor, come to sheep-sit for a week in the Berkshires while the owners of this little farmstead escape to sunny California, where blizzards are as legendary, unfamiliar, and unwelcome as dragons. I prefer New England. In this corner of the world, unlike in California, there's no need to climb mountains to play in the snow.

Cold Spring Farm stands on both sides of Cold Spring Road, tucked into a woody valley a stone's throw from the Otis town line. This is snow country, where winter can start in late October and end with a last, fragile snowflake floating like a dandelion seed on the warm breezes of May.

I CONFESS THAT years ago during a stint at a weather station where my duties required me to read forecasts over the radio each morning, I always had a fear of saying "het, weavy snow." You know, the kind that strains the muscles in your back when you shovel. Although that particular on-air blunder never occurred, I worried nonetheless.

Winter has always been my favorite season. Whenever a flurry of white crystals tumbles and glides from the sky, it cools the ground and refrigerates the air, but never fails to warm my mood.

Snowfall does not limit its range to the North Country, to places like New Hampshire, Minnesota, or Saskatchewan, where winter weather is often cold enough to make a polar bear purr. Sometimes the jet stream sags and droops across the middle of the continent, plunging a wedge of sharp Canadian air all the way down to Texas and the Gulf of Mexico. Unexpected frost zips into Florida and kills all the oranges.

Snowflakes were once spotted over Miami, though the long arm of winter has never yet reached so far south as Key West. Snow can even fall when the temperature warms above the freezing point of water. A layer of cool air aloft releases snowflakes which parachute down toward a warmer layer at the surface; each ice crystal melts slowly as it drifts toward land. A few snowflakes may reach the surface intact, then drip harmlessly into the soil. In New York City, snowflakes were once observed falling while the temperature on the ground held steady at 47 degrees Fahrenheit.

Snow is born in the cold breath of a cloud. Water vapor condenses and freezes around a microscopic nucleus of dust or sea salt carried high on the breeze. In the core of every snowflake lies a tiny speck of atmospheric debris. That, at least, is what the science books will tell you about snow. But folklore has this to say: "When cows don't give milk, expect stormy and cold weather." Also: "When hens run about acting frightened, a windstorm is coming."

The arrival of snow clouds or freezing rain accompanies a steep drop in atmospheric pressure, which is thought to energize and animate animals. Perhaps there's a bit of truth to the matter. Scents and subtle smells undetectable to the human olfactory system permeate the air, floating on the breeze. We happily ignore them all, but the typical dog's nose can follow these odors like trails of bread crumbs. How many hundreds of wild creatures—skunks, feral cats, raccoons, moose, deer—must stalk and prowl through the backyards of America when no one is looking? On fair weather days, high pressure—a convergence of upper-level winds combined with divergence at ground level—will cause air to sink, pushing these odors into the soil. But the advance of stormy weather brings low pressure; the air rises and cools as clouds form overhead. When the air rises, scents emerge. They crisscross the air in loops and swirls, so that a dog's nose must sniff and spin in all directions, trying to identify each new odor. A dog that may have been lying asleep next to a fire suddenly dashes across the yard. The sky darkens with clouds.

American folklore and history is rich with images and explanations of wintry weather. Centuries ago, children recited a curious chant on the morning of February 2, midway through winter: "Candlemas Day! Candlemas Day! Half our fire and half our hay!" The meaning of the rhyme is clear: half the farm's fuel and hay are gone, and half still in stock—at least, it had better be.

In recent years, Candlemas has been forgotten, and the public turns instead to the questionable forecasting abilities of a groundhog in Punxsutawney, Pennsylvania. The unfortunate beast is dragged out of his den and proudly displayed in front of news cameras once a year; he would probably rather be hibernating. How Candlemas evolved into Groundhog Day is unclear, but the day—halfway through the winter season—is still important.

This morning in the Berkshires, no groundhog pokes his head up through the snow as I step briefly onto the porch and brace against the wind. In any case, no groundhog could possibly see a shadow in this blizzard; he'd be lucky to find his way back to his hole. The sky is overcast and swirling. I must rely on a different sort of animal to predict the weather, if I wish to test its accuracy against the official forecast.

Fortunately there are plenty of animals close at hand. The distant barn, invisible in the embrace of whirling snow, is full of hungry sheep—pregnant ewes—who expect me to feed them each morning. But sheep are not interested in weather forecasts, El Niño, or debates on global warming. In my experience, sheep are interested in little else but food.

When I listen for their *baas* from the porch, I hear only wind. Snow blows into my eyes. I turn away.

Officially, a blizzard requires windspeeds of 35 mph or more, plus poor visibility and falling or blowing snow. One storm that perfectly fit that description would later bury the observatory on Mount Washington in 1999—26 inches of snow would fall in a single day, with nearly half that total accumulating in a six-hour period. Sarah Curtis tells the story: "We did *a lot* of shoveling out front. Winds were from the southeast, which is horrendous for us because it comes in everywhere. It took me fifteen minutes to secure a two-by-four against the Dutch door so that it wouldn't fly open. I couldn't shut it anymore. The snow inside the door was probably four feet deep." This morning's weather here in the Berkshires is not so extreme. Wind pushes against the trees and stirs up geysers of blowing snow, but its full rage finally drops below the threshold of 35 mph. Only the smallest branches still tremble in the breeze; the storm toys with the land the way a cat paws its victim before the kill. Perhaps I can be excused for exaggerating and calling this storm a blizzard.

Back in the kitchen, two dogs plod across the cold floor to greet me. Callie, a little black-and-white border collie, yelps in delight and nudges the door with her nose, eager to start morning chores. I grab my gloves and push the door open a crack, flinching against the icy assault of winter air. My nose stings, and a gust flutters the hood of my jacket like a sail. Undaunted, Callie dives off the porch and sprints in an energetic arc toward the barn.

The other dog, Peggy, is a big black lapdog; she has no use for sheep, snow, or freezing winds. She pokes her head through the open door and sniffs once, then stares up at me, pleading with round, dark eyes—surely I don't expect her to go out *there*? Peggy blinks, dismissing me, and waddles back to sprawl across her favorite rug in the kitchen.

When I run outside after Callie, thick mist wraps around me. The snow drapes a scarf of white powder across my shoulders. My breath freezes, clinging in sticky white lumps to the collar of my coat. As I walk, this necklace of ice licks my chin with a cold tongue; I nudge it away with a wool glove.

"Callie," I call. "Callie! Where's the barn?" The dog barks once, sounding puzzled and lost. But at last, the snow starts to ease. The whiteout thins, bleached by the dawn. Horizontal sunbeams skim and skip across the ground, penetrating the fog with light. Sheets of reddish-gold light ripple in the sky. The glint of a billion tiny suns now sparkles atop the drifts, as if Mother Nature has just scooped up a handful of liquid sun and is striding across the meadow with a cupped hand, sprinkling droplets of gold on the snow.

In front of me, abruptly, a barn materializes in the haze.

FARMERS PROVIDE THE best gems of weather folklore. By necessity, they have kept a close eye on weather for many millennia. In the absence of accurate forecasts, they learned to read signs in the turn of the leaves, in the scurryings of ants and other insects, in the swish of a cow's tail. Much of the folklore passed down to younger generations was nonsense—some, perhaps, contains a kernel of truth.

In farming communities in Italy, people once believed that you could avoid catching malaria by sleeping with a pig. How that fact became established, I'm not sure I want to know. Was trial and error involved? Did they first experiment with sheep and other farm animals to determine which species was most effective? Oddly enough, however, this particular piece of folk wisdom is true. Malaria is a scourge brought by mosquitoes, those tiny buzzing seekers of blood who navigate to the next source of protein by following the scent of body heat and sweat. A pig's body temperature is warmer than a human being's, so the mosquitoes will attack the pig first and—in theory—leave a sleeping human in peace.

In addition to fending off mosquito-borne diseases, pigs can help predict the weather. "When pigs carry sticks, the clouds will play tricks," according to one old saying. But no pigs live on this farm, so I must base my weather predictions on the antics of sheep instead.

Outside is a blizzard, yet no sheep shuffle or bleat. The silence is eerie. By now the hungry animals must surely have heard me; the barn door slides open in my hand like a rumble of thunder. Why are they silent? There is one old ewe in particular who always announces my arrival with a metallic *baa-aa-aa,* an impatient sound like the rat-a-tat-tat of an old engine turning over. Today she is silent.

I tug the barn door fully open, digging a trench in fresh, powdery snow. Inside, the air is motionless, murky and dim. All the sheep are gone. The pen is empty. My heart lurches. I turn my head stupidly right and left, up and down. Did I expect them to be hiding in the attic?

A blue gate separates the ewes' shed from the road, but now I notice a frayed strand of bailing twine drooping off the fence, its knot snapped. Apparently a clever sheep has nudged the gate off its hinges and pushed it ajar.

A clever sheep—whoever heard of such a thing? Callie darts ahead of me, slipping through the door behind the shed. She returns immediately—no sheep that way. "How long ago did they escape?" I ask Callie, not really expecting an answer. She barks, a warm burst of noise and enthusiasm that cuts through the cold. There's a glint of energy in Callie's eyes; morning chores are rarely so exiting. She sits down on the hay, tongue awag, staring up at me. I think she's enjoying this.

In the back of my mind I struggle to remember more lines of folklore. "Before a storm sheep frisk, leap, and butt each other." Does that explain the frayed twine, the broken gate—playful antics before the storm?

Another aphorism informs me, "When sheep go to the hills and scatter, expect nice weather." At the moment, that's not very helpful.

A joke from a century ago highlights the flaws of relying on farm animals for weather forecasts. I first encountered it in an out-of-print book called *Weather Wisdom,* by Albert Lee. According to the tale, when the local weather predictor in Springfield, Illinois, failed time after time to warn his neighbors of coming storms, the mayor of the town began to grumble. Then one day, a young boy claimed he knew when a storm was coming—every time. The mayor, somewhat skeptical, gave the boy a chance anyway. Quickly the boy demonstrated his accuracy, foretelling the arrival of several storms. So the mayor decided to get rid of his old weatherman and hire the boy instead. But to his surprise, the boy refused the honor. "Mr. Mayor," he said, "it's my jackass that does the predicting. Whenever a storm is buildin' he scratches his ear on the fence and brays something awful."

So the mayor appointed the jackass instead.

According to Albert Lee, the sixteenth president of the United States, Abraham Lincoln, added his own punch line to the joke. He muttered that giving that job to the ass was the mayor's biggest mistake. Why? "Because ever since then jackasses have been seeking public offices."

Ever since then, weather watchers have been blowing forecasts, too—particularly when it comes to snow. How many times have you heard a local meteorologist call for 15 inches of powder, only to receive a dusting, or even a cold rain?

"It's a very difficult thing to forecast, but that's what people want to hear about," says Sarah Curtis at the Mount Washington Observatory. She shakes her head. To prepare a forecast—in any season—you first look at the current radar, satellite, and surface maps. You check where the fronts are, the centers of highs and lows, and any surface observations. "I'd look at all that just to get an overall picture of what's going on in the atmosphere," Sarah continues. "Where the warming temperatures are, where the cooler ones are. To forecast when and how the systems will be moving through, I'd use the forecast models: the NGM, the Eta." The acronym NGM stands for Nested Grid Model. It is a computer simulation of the atmosphere, and is run twice a day, at 0000 Greenwich mean time and again at 1200 GMT, offering predictions up to 48 hours in advance. For a less certain, longer-range forecast, the one from the ECMWF (European Center

for Medium-Range Weather Forecasting)—is an option. It offers predictions up to 240 hours in advance. With computer models like
these, technicians simply toss in all the data—500 millibar measurements, 850 millibar, surface observations, et cetera—and the computer analyzes what will happen. However, the use of computer models
doesn't mean instinct and experience no longer play roles. In the case
of TV meteorologists, "they're all using the same products. They all get
the DIFAX charts from the National Weather Service," Sarah explains. DIFAX stands for Digital Facsimile Service. "That's the main
tool, or something similar to that. I think the difference you get is from
the instinct and from people's different experiences. I tend to lean
toward people who've been in the business the longest." She gestures
with her hands to emphasize this final point. "I don't want that to
sound like age discrimination or anything like that, but you truly learn
by being in this business as long as you can, and experiencing things,
instead of just reading the models. Because if the models were perfect, we wouldn't really need meteorologists to interpret them and try
to figure them out for everybody else."

THE ONLY THING I must predict at the moment is the behavior of
sheep. How will I find them? After a moment of thought, I walk over to
the feed bin and yank open its lid. So much for folklore; I have discovered a more practical way to catch sheep in a blizzard. A thin odor
of wheat rises from the feed bin to tickle my nostrils before dying in
the cold air. With a large scoop I fill a bucket with feed—sheep-bait.

Outside, the snowy ground lacks even a single hoofprint. Wind
has erased the tracks. But Callie now trots ahead and barks once, as if
to say "Aha!"

Not far from the edge of the road we find a patch of droppings,
half-buried by drifts. Ghostly steam rises off the manure. A string of
fresh hoofprints leads to the road—but there they vanish again. Did
the sheep wander down the road?

I still have a plan. I shake the bucket, a raspy rattle of grain pellets. Shake, shake, shake, shake—and about one minute later, the

missing flock emerges from the trees. They dash toward me, a stampede of wet wool, eyes fixed on the bucket in my hand. Newly fallen snow melts on their backs and mats the wool. I brace for their sudden assault. An ewe's nose nudges my knee. A dozen more sheep brush against my legs.

Slowly, I wade through this logjam of sheep backs, luring them back to the shed. Callie sprints across the snow, helping me keep them all in tow.

Inside, I pour out the feed, and soon the tight clamp of sheep bodies eases around my legs. Snow melts off my shoulders and drips to the floor. I peel twine off a bale of alfalfa and dump it in the feeder.

Callie whimpers and pokes her nose through the gate; she hasn't had enough fun yet. As I loop a chain over the gate, it clangs loudly on the bars. A town plow rumbles by on Cold Spring Road. Its engine whines as it whips past me, then vanishes down the road.

In the sky, the wind scoops up funnels of loose, dry snowflakes. A coil of blizzard snaps like a whip above the silo, then crumbles, falling back to earth as a gentle powder.

The sheep's bleating stops abruptly, and a new sound reaches my ears. Echoing off the walls, magnified by the chill crystalline air, I hear the crunch-crunch-crunch of sheep teeth, busily grinding their breakfast of frosty alfalfa into a warm, viscous, digestible soup of starch.

Even in the crystal cold, chilled by snow and wind, I stop a moment to watch the sheep. I shiver and cross my arms, standing shin-deep in a snowdrift at the edge of the barn. The sheep appear content, immune to the probing fingers of the wind. There's another old saying, this one from the Mediterranean: "With good wind, good bread, and peace at home, if snow does come, let it come."

Let it come. I'll appreciate the fireplace all the more after this excursion in the cold. In my opinion, winter isn't so bad—in fact, it's the best season of all.

10

Snow and Ice

IN HIS ESSAY "THE WIND," AUTHOR RICK KEMPA QUOTES the folklore of ancient China: "There is no wall through which wind cannot pass."

New Englanders speak of wind in equally dire tones. "No weather's ill if the wind be still." And also: "When wind is in the east, it's neither good for man nor beast." But if a ridge of high pressure builds south from Canada on a clear February night, and the air temperature plummets to ten below zero as a swift flow pushes chimney smoke horizontally, it scarcely matters anymore which way the wind is blowing. You can say whatever you want about the weather—but your teeth will be chattering as you say it.

Tonight the hour is late, and winter's breath penetrates walls and windows with ease. A paper-thin gap under the front door is an open invitation for icy drafts. No crack or crevice is too small to escape the wind's attention; air molecules infiltrate each one. Even the place where I am sitting in the living room provides no refuge, despite a barrier of storm windows and insulated walls. A cold current ripples across the floor.

The fire in the wood stove crackles and sparks, open in the front except for a black screen. It is the only source of heat. Bubbles of energized air rise off the cast-iron surface, buoyant and light, uselessly warming the uninhabited spaces near the ceiling. Air floats away from the replenishing heat of the stove, surges to the far wall, cools and sinks. Cold air is denser and therefore drops to the floor, undulating over my toes back to the stove, where it is reheated and rises. The cycle starts all over again.

The dichotomy of hot and cold plays tricks with the walls and floors; they expand and contract as the temperature changes. I hear sudden footsteps creak and moan on wooden planks behind me. But no one is there. An unexpected paw of wind reaches over my shoulder and swats playfully at the flames. The fire shrinks back for an instant; briefly, it casts no warmth.

Trembling, I rise to fetch a sweater, an extra layer of insulation. My body heat leaches out through toes and slippers, disappearing into the wood. Goose pimples form on my skin, tiny mountains of discomfort rising like the Himalayas up and down the length of my arms.

When I step close to the fire again, the flames cast shadows against the far wall, dancing in silence. Vines of frost grow and cluster on the glass windows. Water vapor in the air flows invisibly until it rubs up against the chilled pane; there, it freezes on contact into microscopic bubbles of ice.

I look outdoors and see the shadow of a snowflake in moonlight flit across the sky. Jack Frost presses his face against the glass and peers back inside.

SLEEP ELUDES ME. I carry a book and a mug of cocoa close to the fire. The radio plays jazz until interrupted by a commercial for antifreeze.

A deep voice on the radio describes New England as "the kind of place where people talk fast, so their words don't get frozen solid." I chuckle at the quip. Except for the monaural chatter of the AM radio, the entire house is silent. No one else is here, and it is too cold to talk in any case. So I listen.

Three days ago a noteworthy blizzard struck the Midwest, knocking out electrical power in portions of several states and paralyzing the roads. Snowdrifts buried the interstate highways, stranding thousands of people in their cars; they huddled together, waiting for rescue while their fuel gauges tipped toward empty.

Years ago, I remember listening to another news flash about a virtually identical late-winter storm on a radio talk show. It was early March of 1997. Frequent bursts of static interrupted the announcer's tense, worried voice. At one point, a police officer whose cruiser was stuck in the snow phoned the station with grim news. He carefully enunciated each word: "Even if an emergency crew—police or ambulance—had to get through, the lanes are blocked. There's no food or water. We may have to get the National Guard to airlift."

Another caller from a cellular phone worried aloud. "There's thousands and thousands of cars. I've been here hours and haven't moved, not at all. Even the trains are frozen in. The O'Hare airport is closed. It's like a holocaust out there."

"The road's impassable. People are running low on gas. There's a lot of people in the same predicament. All we can do is grin and bear it."

"They've called electricians in all the way from Texas to help with this catastrophe."

The current storm follows the same pattern. Earlier in the week, the snowplows were busy in Chicago. Now it is three days later, and what's left of the storm is on my doorstep. I turn off the radio and reflect on the fact that weather patterns move west to east; the storm is scheduled to arrive by daybreak, perhaps sooner. The passage of 72 hours has barely weakened its intensity. I check the porch for a shovel—it's there.

A sudden gust of Arctic air rattles the windows; the lights flicker and brown. Icicles plummet off the roof, landing in brittle, glassy chunks on the porch floor. Morning is close, but the weather forecast says it will bring little warmth and much snow.

WINTER'S VAPORS CAN kill. In March 1997, eight people died during that severe winter frenzy that pounced so unexpectedly on the

Midwest. But here in New England, Old Man Winter is usually subtle and sly. He charms us by painting pretty pictures in the windows,
then waits assiduously for the fire to burn low. I toss on another log to
deny him the opportunity.

Snow and ice kill more people in the United States and Canada
each year than hurricanes, droughts, earthquakes, or wars. Yet no one
truly fears winter. An overnight blizzard sculpts insurmountable white
dunes on the city streets, but seldom elicits more from us than a
grumble. We shovel out our driveways the next day and think no more
on the matter. Children rejoice and cheer at a heavy snowfall, because
school is closed for the day. Mother Nature has outdone the principal.
Skiers and snowshoers take a holiday.

I remember from childhood the pleasure of digging deep caves in
snowbanks, tunnels in which to store an arsenal of mitten-packed
snowballs for the harmless neighborhood battles that all children
wage.

I once owned a red sled upon which I hurtled down a steep hill
at the edge of the cow field behind school. As I grew older, only half
my body fit on the sled; I dragged my feet behind me to brake the icy
free-fall before slamming into the fence at the bottom. The cows
watched dully. I was so young and insouciant that I don't even remember wet snow spilling into my boots, or the long trudge back uphill, or the burning sensation in my chest as frigid air seeped deep into
my lungs.

What I do remember is a family tradition—shaking the snow off
a Christmas tree deep in the woods on Christmas Eve and bringing it
home. Our family paid for the privilege by handing two dollars to one
of the two elderly German sisters who lived above a gurgling river in
the spruce grove. The financial transaction was legal, paid in advance,
and perfectly mundane. But in my mind we always slipped through
the low afternoon shadows like poachers of trees, saws in hand. Shadows lengthened. A waxing moon glimmered somewhere in that dark
sky, but its silver light splintered and blurred in the midst of billions of
falling snowflakes. The whole sky shimmered faintly; we couldn't see
the source.

White mounds of snow on the ground blushed a pale red as the sun quickly set. Falling snowflakes ignited with sudden color, like sparks and embers shaken loose from the sky. When a suitable tree finally appeared, I wriggled under the dark branches; the spruce needles raked across my coat. My scarf caught in a snag from which I pulled free, releasing a cascade of cold ice crystals from the upper branches.

My early memories of snow are always good ones. Not until I had graduated from college and was house-sitting on a farm in February did I fully realize that winter, while beautiful, never hesitates to kill. An obituary caught my eye; it was on a crumpled, coffee-stained newspaper that I was about to use to reignite the fire. I hesitated. It documented the entire 82-year life of an elderly neighbor in four succinct paragraphs. He died of a heart attack, clutching the handle of a shovel in his driveway with bone-white fingers. I skimmed the page, a summary of the life of a neighbor I had hardly known. Then I pitched the paper into the flames; it blackened into crisp feathers of carbon.

The dry, fluffy powder that often falls in Colorado and other western states cannot match the leaden weight of Eastern snow on the end of a shovel. The exertion of clearing winter's debris from our driveways claims hundreds of lives each year.

Accidents on slippery roads lead to further casualties. "It's not myself I'm worried about," a friend of mine once said while navigating a white road in his car. "It's other people I worry about." The technique of braking and stopping on ice he knows well. But even the most experienced, cautious driver can do little to avoid collisions with reckless or less skilled motorists.

Nonetheless, winter had always been my favorite season. So why was I so surprised to discover the season's danger, its violence? After all, one of my earliest memories of winter was the night the wind savagely ripped apart a willow tree in the backyard. The sky had hissed and screamed like a cougar as gusts sunk their bitterly cold teeth into the wood. I heard the tree moan. By morning, the sky had fallen silent. When I peeked out the window, shading my eyes to block out the glare of sunlight reflected off the snow, the tree was gone. Uprooted? Devoured? Splinters of bark, black sticks, and twigs stabbed into the snow drifts like porcupine quills.

WITH ONE FOOT I stuff the carpet closer to the doorway—a futile effort to keep January air outside where it belongs. I rub a hand over my arm to generate friction and warmth, but the pinprick mountains of goose pimples—technically, the term is piloerection—stay stubbornly in place.

Goose bumps, gooseflesh—whatever we call the phenomenon, its manifestation on arms and legs bears a distinct resemblance to the bumpy skin of a plucked goose or chicken. Goose bumps are a biological relic of our mammalian ancestry. Hairs on a mammal's hide always arch and curve in cold air. On bears, the thick mat of hair bunches; it traps bubbles of warm air in tiny cavities under the brown fur. The trapped air acts as an insulator close to the skin, a million tiny bubbles of warmth rolling up and down the body.

That process works well if you are hirsute as a bear. But for humans, who long ago shed the shaggy hides of their primate ancestors, goose pimples are all we have left. They are useless; the warm air simply leaks away. So we must substitute wool sweaters and padded coats for the furs of our evolutionary past.

Darkness and cold seep through the windows early in the winter months. The tilt of Earth's axis guarantees a short day. At the North Pole, the sun refuses to rise for six months and Santa Claus groans at his oil bill. Down in the middle latitudes of New England, twilight grips the sky as early as 4 P.M.—an hour that was once considered midafternoon, back in the warm dream of July. Now the night seizes the sky swiftly and hard, pilfering warmth and color. Stars shine like crystal ice, distant and aloof. The air stills, chilled. Beyond the red halo of the fire, it feels too cold to move. Deathly cold. My imagination soon gets the better of me, and I speculate about what might happen if the submicroscopic air molecules themselves were suddenly frozen in place. What if the temperature dropped to the impossible limit of absolute zero, 459.67 degrees below zero on the Fahrenheit scale? At that point, even atoms stiffen and go numb. Gusts of wind hang like icicles. The air waits in anticipation of dawn.

"AS THE DAYS lengthen, so the cold strengthens," is an old proverb.

In winter, days actually grow longer here in the North Country, allowing us ever more exposure to the sun's warm rays. But January and February are still the coldest months of the year, and by far the most savage. Wind gnaws at the snowdrifts and shatters icicles with loud growls. Listening, I can almost imagine a great beast hunched outside the door, claws groping around the house trying to rip the basement out by its roots.

Meteorological winter starts on December 1, though most calendars place the official first day of the season on the shortest day of the year, December 21—the winter solstice. Beyond that date, the days start to grow longer by one or two minutes per 24-hour period. But they are still much shorter than the nights. As a result, the ground spends more time cooling than heating, a trend that continues until the vernal equinox in late March, when day and night temporarily equalize at 12 hours apiece. Until then, the air stays boreal and fierce. The chilling howls of bitter winter winds can only be assuaged by the comfortable crackle of the fireplace.

"February is only good for filling ditches," according to weather folklore. But I wonder. In the apocryphal Good Old Days, winter wasn't always viewed as harsh and unwelcome. Our modern technology has provided us with wintertime problems that our ancestors never faced. For instance, in the nineteenth century and earlier, no one's wagon ever refused to start in the cold. Nobody worried about the price of electricity or heating oil. Snowfall actually made travel easier; as lakes and rivers froze solid, a number of flat, smooth, natural bridges solidified above the water. Imagine a wooden-wheeled wagon with no shock absorbers lurching and bouncing over a muddy road in summer and autumn: by comparison, the quick glide of a one-horse sleigh on snow surely delighted weary travelers.

My great-aunt once told me about the joys of skating across frozen ponds in Newfoundland in the early 1900s. She also described the beauty of the frozen ocean off the coast; the sea's flattened waves petrified like a snapshot of a moment.

Although the woes of transportation actually lessened during those long-ago winters, the temperature stayed mercilessly cold. Average worldwide temperatures dropped by several degrees between the years 1450 and 1850 A.D., a period known as the Little Ice Age. Since people couldn't just crank up the thermostat or fly to Florida, they invented other ways to keep warm. The old expression "firewood warms you twice"—once when you split it and again when it burns—probably appeared in the lexicon during this era.

In New York and New England in the eighteenth century, colonial Americans didn't have polypropylene pajamas, so they invented a way to keep warm and find a future spouse at the same time. They called it "bundling." Rather than send a suitor home—a long and difficult journey on a frigid winter night—his fiancée's family often invited him to spend the night in the arms of his belle. Naturally, bundling raised a few hackles in puritan New England. Sometimes anxious parents attached bells to the bed to make sure the young couple shared nothing but pillow talk while fending off the chill with body warmth.

Modern civilization finds itself inconvenienced by winter weather to a greater degree, because there is always more going on, more to do, more to disrupt. Even as far back as 1888, in the modernized city of New York, an unexpected ice storm in March sagged power lines and telegraph poles and shut down the stock exchange. The wind piled snowdrifts over second-story windows. More than 200 people died.

MORNING ARRIVES AT last but brings little warmth. My car engine whines and coughs as if it caught a cold overnight. "Late again," I grumble with a glance at my watch, which is a cold metallic band around my wrist. A few more cranks of the key prompt a sneeze of exhaust; the engine finally catches. I tap a foot on the gas pedal and peer ahead at the road.

I am not the only early riser. Far down the street, I see a man in rubber boots slip and topple in a pile of flailing arms and legs on an icy patch next to his mailbox. The edges of his black leather coat flap like pterodactyl wings. Slowly he rises, first to his knees, then to his feet, and I see his lips move as he mutters to himself. He takes a moment to straighten his wool hat. Then with an angry hand he swipes an icicle dangling from his mailbox. It shatters on a clear patch of pavement like brittle glass.

Ten inches of snow have tumbled from the sky overnight and in the early hours after sunrise. The road has vanished. I squint through a hole in the frost on the windshield; the side windows fog with moisture and quickly turn opaque. Jack Frost seems determined to blind me. An icy vapor seeps through the car's vents until at last the engine warms and the defroster kicks alive, huffing warmly. A spray of heat shoots through the vent and excavates a small circle of clear glass on the windshield.

I shiver, huddled deep inside my coat. My jacket is a portable replica of bear fur, an artificial hide which I eagerly shed in summer but must depend on for survival during the winter months.

The car's side windows finally clear a bit as the interior warms. Along one side of the road, white sleeves of snow clothe the branches of evergreens; they dip and sag, reaching to the ground. Smaller mitts of snow dangle from the tips of twigs like loose white gloves.

All snowflakes develop in a six-sided pattern, as if sculpted by the same artist. Every schoolchild learns the adage that no two snowflakes are exactly alike. But that is only half the story. Air temperature is what determines the eventual size and shape of snowflakes: needles, cones, or plates are all possibilities. Temperatures in the low 20s Fahrenheit produce sharp needles of snow. Warmer temperatures create hexagonal plates of ice that glide softly to the ground. The classic snowflake is simply an aggregate of many ice crystals, often forming in crisp air at around ten degrees Fahrenheit.

Varieties of snow come with many names: the Eskimos are said to call snow *qanik* as it falls and *aput* once it settles on the ground, draping the landscape with a sheet of white ice. Drifting snow, actively funneled and piled into heaps by the wind, bears a tongue twister of a name: *piqsorpoq*.

This morning the birch trees bend and stoop to ground level. The wind scoops a wet clump off the end of a twig; it lands with a splat on my windshield. The sky hangs low, a gray roof of mist. No sun or moon exists. Tumbling snowflakes materialize out of nowhere just inches overhead—or so it appears. I cannot see their source. The flakes do not seem to drift downward; instead, the ground rises to meet them. All reference points are obscured by the ice-cold haze.

Town plows have not yet reached this rural back road, so my car's tires dig ruts in a white river of crystalline ice. Except for the winding tracks of the single car to precede me, the road is invisible. My tires slip and spin for an instant; gravity loosens its grip. Then all at once the ground clamps down again and I have traction.

Soon I come to the hill—The Hill, I should call it—and suddenly all bets are off.

The nose of the car lurches down like an airplane with no lift. The tires accelerate. I gently tap the brakes, then floor the brake pedal in desperation. My car takes no notice. It continues plunging down the hill toward a pair of railroad tracks. No matter what I try, the frictionless road refuses to acknowledge the presence of my tires. My stomach rises in my chest; I am essentially diving through air without a parachute, and the ground below is rising in a blur to squash me flat. The steering wheel cuts limp and wags from side to side. Tires rotate and whir without direction.

Only 100 yards away, the railroad tracks intersect the road. What worries me—surprisingly, I have time to worry—are the two cement posts that jut from the ground at the edge. They intrude into the center of my vision. For an eternity, the cement barriers loom ever larger but never quite arrive. Then time kicks back into gear. The cement posts make a beeline toward the hood of my car and I don't even have time to shut my eyes and pray.

Snow is simply one of several causes of winter automobile accidents. Freezing rain is another. This seems an appropriate time to discuss the reasons why.

Freezing rain starts as a snowflake tumbling from the sky. If a middle layer of warm air waits below it, it will melt, only to cool again in the cold air close to the surface. Large, supercooled drops can exist as a liquid well below the normal freezing point of water—down to minus-40 degrees, in fact. Then they hit solid ground and freeze on impact into a hard sheen of ice.

Glaze is glass, a rock made of water, hard as stone until the sun's caress makes it drip away like butter. Raindrops ride the wind and plummet. The drops splat onto pavement, streetlights, power lines, trees, or cars, and harden into ice. Mere snow yields like powder to the blunt edge of shovels and plows, and the delicate feathers of rime ice may be removed with a slap, but glaze is far more stubborn. Glaze is ice you must mine; only a pickax and sweat can clear the way. Or time. Pour dollops of warm sunlight on the ice—or simply wait for winter to turn to spring—and the glaze finally drips into the ground, a harmless dew.

In January 1998, a severe ice storm knocked northern New England and sections of Canada back into the Stone Age. Power lines sagged; power stations crumpled. A so-called Triangle of Darkness engulfed southeastern Quebec and much of northern Maine.

"A strong El Niño winter produces the conditions that are right for an ice storm," explains Dave Thurlow safely after the fact. His own home in New Hampshire escaped with only a few hours of no electricity. Other nearby communities fared far worse.

Thurlow takes a moment to describe the structure of the atmosphere at the time of the storm: "You have discrete layers in the atmosphere that are just above and just below freezing. In a typical ice storm, you could have a cold layer at the surface, and then a warm layer above. You could even have intervening layers of cold, then warm."

The ice storm of 1998 was not an especially strong storm system, he points out. "It just happened to last a long time. And the layers of air happened to hover just above or below freezing. Ice storms have virtually no wind. Their precipitation isn't necessarily strong, just continuous."

Storm systems usually travel along the path of the jet stream. In January, however, the storm stayed put. "It was exceptionally slow," says Thurlow. A look back at the weather maps reveals a long, stationary front extending across Quebec through New England into New Brunswick. "The jet stream was parallel to the front instead of perpendicular. If it was perpendicular, it would have moved that front right out of here."

No matter how much they complain, New Englanders pride themselves on surviving bitter winter weather. The relationship with the weather is give-and-take. Generations ago, ice storms and blizzards were thought to build up character, though they also tore down roofs and destroyed trees. But the ice storm of 1998 proved almost too much even by hardened Yankee standards. Rescue and repair efforts lasted weeks, and in that time I overheard a conversation between two men from New Hampshire and Maine.

"You know, whenever there's a catastrophe here, it really brings out the best in people. People pull together, work together, try to get everything up and running again. I was down in Florida during a hurricane and the people didn't do nothing." The man practically shouts the last word. "They just sat on a curb and waited for someone to clean up the mess."

"The federal government, I guess."

"It's a matter of survival. Down there, it's warm. You can survive just waiting around. But if you're up in Aroostook County in winter and you see a car on the side of the road, you better stop to help."

I CAN ONLY hope someone stops to help me this morning. The clouds hang low, blotting out the sun. Globules of cumulus fractus fray in the wind, like cotton balls pulled apart. A thick stratus layer blankets the hills. No warm rays penetrate from the sky to help chip away the ice. And I do not have the luxury of waiting for spring.

I have just lost all friction on the steepest hill in the county and am careening out of control. But at least I now know the cause. A clear sheet of glaze is disguised beneath the snow—a delicate trap, set overnight. It cannot be plowed. Even with sand on the road, it will not melt in the second or two I have left before striking the bottom. My car lurches and slides, unstoppable.

I wrench the steering wheel to the left. A second passes, then two. Three seconds. I brace for impact. The cement posts streak close. My vision blurs.

Miraculously, I rush past the half-seen shape of one post with perhaps an inch to spare. It happens so fast that my brain hasn't caught up; I don't quite realize yet that I'm still alive and uninjured. Instead, I breathe, gulping. Finally, I can ease the tension in my knotted muscles.

The slope of the road flattens ahead, but still my car veers into the far left lane. The hood points directly into the trees, where thick branches full of snow wait to catch me.

I wrench the wheel back to the right, and at last the tires strike pavement. I zigzag across the road in a long wide "S" for half a mile, powered only by the momentum imparted to the mass of my car by the hill now fading in my rearview mirror. I see a white van appear and ease down the hill more slowly behind me. The driver probably witnessed my tortuous descent.

Only now do I have time to think, to exhale, to worry in retrospect. What if a train had thundered across the tracks at that moment? What if my car had hit one of the posts, so that I helplessly watched the hood crumple until my head snapped against the windshield and the movie of life spun off its reel? Or what if another vehicle had driven toward me while I spun so wildly in the wrong lane? Another casualty of winter would have been written up in the obituaries. Maybe someday in a cold living room, a stranger would have glimpsed my name in the paper before crumpling the page into a ball and tossing it onto the flame.

Winter plays no favorites. I drive on, the car now steady but my hands trembling on the wheel. My heart pumps panicked blood and adrenaline through my body. I feel a fire, a warmth that the ancient mammalian goose-pimple effect had failed to provide. Though still afraid, at least I am no longer cold.

Cold Spring

In May comes the tail of winter," according to an old farmer's saying from France. "Till April's dead, change not a thread," commands another maxim—meaning it's best not to stow your winter coat in the back of the closet just yet.

Springtime is when the white cocoon of winter warms and peels; southerly winds flush the first grass green. The jet stream slides up into Canada, allowing warmer air from the Gulf of Mexico to push into the north. On the fingertips of twigs, buds sprout copiously in the trees. Overnight, a sparrow's nest is stitched to a branch on a silver maple. Honeybees start to buzz and hum in the ecstasy of the new season's pollen; the first flowers bloom.

Crickets once again leap through tall grasses, mindlessly singing weather reports to all who listen. The chirps and squeaks of crickets hold a special meaning; the rate varies as air warms or cools. The body of a cricket, as it noisily rubs its legs together, provides the curious naturalist with a perfectly precise thermometer. Count the number of chirps produced by a single cricket for 15 seconds and add 37. The resulting number is the exact temperature in degrees Fahrenheit. It never fails (provided you find the right species of cricket). Folk wisdom is rarely so precise.

I hear no crickets early on this April morning; the spring season is still too young. The air nips with cold. A puddle of melting snow ripples in shadows at the edge of the trees. Pinecones float in the cool liquid like dark wooden shells. The exposed soil in the fields waits in black clumps, moist with the promise of grass and hay yet to come.

At dawn, I step quickly through the crisp air between house and barn. This Massachusetts farmstead where I am working is shaded by trees; the sun hangs low, and cool temperatures linger. The wet residue of last night's frost still glistens on the grass stems, but deep in the forest I hear sparrows and finches sing songs about spring. A breeze dabs at my face, heavy and moist. I must walk through weighted air to get where I am going. A surging ocean of oxygen, nitrogen, and water vapor pushes back against my arms and legs. I must shove it aside with brute force, the way one nudges open a stuck door. A whistling current of wind ripples and eddies around my limbs, and I am entwined.

Overhead, the clouds bulge in gray sacs filled with rainwater and do not move away. The sky is fey, moody. Even though a flicker of sunshine cuts through the overcast and sprinkles the ground with short-lived shadows, the rigid roof of clouds threatens a downpour of cold April rain.

The sky soon purples and warms. Louder breezes start to grumble in the hills. Tall trees—50-foot leafless birches that lean protectively over houses—flail and lash suddenly about the sky. Vaguely, I recall a melodious voice drifting over the radio this morning to warn us of thunderstorms on the way. (The radio itself can provide a warning, issuing a squeal of static on the AM dial due to electricity in the air. A thunderstorm emits booming noises in the range of ten to 300 megahertz, which human ears cannot detect but AM radios can. The sound is higher in pitch during the early stages of thunderstorm development, and drops in pitch as the storm matures.)

A mile north of here, in immediate answer to my thoughts, a sizzle of yellow flame slices across the sky. An explosion ruptures the air. The ground shakes, and chattering birds fall silent. Pellets of hail and rain now rattle against the barn's roof, the cloud ceiling having cracked and given way.

Within minutes the flurry of angry noise dims to a soft patter. The deluge is already over. I recall an old saying about weather: "The sharper the blast, the sooner it's past." This morning's thunderstorm proves the aphorism true. Early American folklore also remarks, "The sudden storm lasts not three hours." Or, in this case, not three minutes.

I walk back outside to feel the breeze on my face. Cool mist clings to my skin. The evaporation of fallen rain chills the air by several degrees. But the base of the storm has already passed by.

A curtain's been pulled aside; the sun gleams. Rainwater drips off branches and plunks into puddles on the ground. Each drop sparkles and splashes like an aqueous diamond. Beyond the hills, I can hear the distant thunderstorm grumble once more and then subside, slipping back to sleep.

The cumulonimbus clouds that produce lightning and thunder can also generate tornadoes if conditions are right. Statistically, certain areas of the United States are more vulnerable than others. "Tornado Alley" reaches from Texas across Oklahoma, Missouri, Kansas, and Ohio, and endures more tornado outbreaks per year than anyplace else in the world. Springtime marks the beginning of a loosely defined "tornado season," which extends into August. The two months with the most frequent twisters are April and May.

Michael Morrison, a cheerful man who fields questions about weather in the Mount Washington Observatory's valley museum shop, remembers watching a tornado devastate Xenia, Ohio, in 1976. "Where we were the air wasn't moving. Heavy, humid," Mike remembers. "But nine miles away there was this *thing*." He accents the last word. "What was incredible was the way it changed color as it went, depending on what it was sucking up. It was white over water, then it would turn green—ripping up grass and leaves, I guess. Then it turned black, and I don't know what it was doing. It was incredible."

Even before the actual funnel develops, clouds will often turn a menacing, dark-green color. This is caused by sunlight passing through enormous amounts of ice—hail—suspended in the air by powerful updrafts. As a result, the feeble rays of light that manage to penetrate the seven-mile-high cumulonimbus cloud acquire a bluish-green tint on the way down, which is sometimes enhanced by the reflection of foliage at ground level off the base of the cloud.

On Palm Sunday in 1965, a swarm of tornadoes—37 twisters in all—swept through sleepy Midwestern towns, injuring more than 5,000 people in one of the deadliest tornado outbreaks ever. An equally ferocious outbreak of 78 tornadoes tore through Oklahoma, Kansas, and Tennessee in the spring of 1999, killing dozens and leaving thousands more homeless. One of the biggest funnels cut a swath of destruction half a mile wide at its base, with radar-clocked winds faster than 300 mph.

Jacob Klee, a weather forecaster in Oklahoma City at the time, escaped the worst of the 1999 outbreak, though parts of the city were in ruins by the next morning. "The first reports that came in were for four-and-a-half-inch hail in southwest Oklahoma," he recalls. "I really don't remember when the first funnel was sighted, but it was just a little rope to the best of my knowledge."

The "little rope" quickly enlarged, and the situation steadily got worse. "In the office, while the tornado was passing through the city, our power was flickering like you wouldn't believe," Jake tells me in his usual rapid-fire way (hence his nickname, "Jetstream").

"At first the storm was on a track which would have taken the tornado just a few miles to the east of where I was working, and even closer to my apartment. As it turned out, the tornado passed between ten and twenty-five miles from us to the south and east. We also had a few other storms which were headed in our direction which fortunately changed their direction enough to stay away, between twenty and thirty miles to our north and west." The cell later shifted north and west, and an F4 tornado nearly a mile wide steamrolled through the little town of Mulhall, Oklahoma. "Those in and around Mulhall lost everything, because the tornado destroyed roughly ninety percent of the town."

He pauses, takes a breath. "The TV stations did not make matters any better by showing unedited video of families desperately searching for loved ones, only to find them among the dead."

A day later Jake tempers that stance with some praise. "As sickened as I was by some of the on-air sensitivity in the heat of the moment, I am impressed with how well the local media did at alerting the region to the impending disaster. In my opinion, it prevented the toll on human life from being much, much greater."

Monster thunderstorms called "supercells" spawned 78 tornadoes that day. The meteorological wrecking ball that slammed through Oklahoma City generated wind speeds measured by Doppler radar at 318 mph, at 150 to 300 feet above the ground, pushing the limits of the F5 classification on the Fujita Tornado Scale. It was the first F5 tornado to hit Oklahoma in more than 17 years, since April 2, 1982. It was the first F5 *ever* to plow through Oklahoma City proper.

In 1971, Theodore Fujita of the University of Chicago developed his now-famous scale, which comes up with estimated wind speeds for tornadoes based on the degree and type of wind damage. His scale was a valuable innovation because, more often than not, tornadoes damaged or destroyed any anemometers directly in their path. The scale ranges from F0, with speeds less than 72 mph, to the strongest, F5, which features maximum winds of 261 to 318 mph. Fujita considered an F6 unlikely ever to occur.

"The whole point of the Fujita scale ceases to exist after F5," explains Jacob Klee.

"Because there's nothing left to destroy?" I ask.

"Exactly. Because the Fujita scale is based on structural damage." Above 300 mph, little or nothing survives.

According to the National Weather Service Forecast Office in Norman, Oklahoma, the May 3 outbreak destroyed two schools, three churches, 85 businesses, 473 apartments, and 1,780 homes in Oklahoma and Cleveland counties alone. Thousands of other structures were damaged.

A public-service statement released by the National Weather Service and National Severe Storms Laboratory on May 6 included the following information:

JUST SOUTH AND EAST OF AMBER . . . THE TORNADO QUICKLY GREW TO CLOSE TO THREE QUARTERS OF A MILE WIDE. AS-PHALT PAVEMENT . . . ABOUT ONE INCH THICK . . . WAS PEELED

FROM A SECTION OF RURAL ROAD . . . ABOUT FIVE MILES EAST OF STATE ROAD 92. THE FIRST DAMAGE RATED AT F4 WAS DISCOVERED ABOUT FOUR MILES EAST-NORTHEAST OF AMBER. F4 DAMAGE WAS OBSERVED CONTINUOUSLY FOR SIX AND ONE-HALF MILES . . . WITH ANOTHER AREA OF F4 DAMAGE ABOUT TWO MILES NORTHWEST OF NEWCASTLE. TWO AREAS OF F5 DAMAGE WERE OBSERVED. THE FIRST WAS IN THE WILLOW CREEK ESTATES . . . A RURAL SUBDIVISION OF MOBILE HOMES AND SOME CONCRETE SLAB HOMES IN BRIDGE CREEK. TWO HOMES WERE FOUND COMPLETELY SWEPT FROM THEIR SLABS . . . AND ABOUT ONE DOZEN AUTOMOBILES WERE CARRIED ABOUT ONE-QUARTER MILE. GRASS VEGETATION IN THIS AREA WAS COMPLETELY SCOURED TO MUD . . . AND SMALL CEDAR TREES WERE LEFT DE-BARKED AND DEVOID OF GREENERY. THE RIDGECREST BAPTIST CHURCH WAS DESTROYED NORTHEAST OF THE FIRST F5 DAMAGE AREA.

THE SECOND F5 DAMAGE WAS ONE MILE WEST OF THE COUNTY LINE IN BRIDGE CREEK . . . AND CONSISTED OF A CLEANLY SWEPT SLAB HOME WITH FOUNDATION ANCHOR BOLTS AND ANOTHER VEHICLE LOFTED ONE QUARTER MILE. THE MAXIMUM WIDTH OF THE TORNADO IN BRIDGE CREEK WAS ABOUT ONE MILE.

CERTAIN ATMOSPHERIC CONDITIONS WERE necessary to establish this powerful supercell. "There's the contrast between the cooler and drier air. You had cool air coming from the northern Rockies and dry air from there and the southwest," says Klee. "Then you had Gulf moisture—warmer, southerly winds. You had high pressure up in the Southeast, which was drawing southerly winds on the back side of the high. Of lesser importance, the high kept the storm moving slowly.

"It was a strong dry line," Jacob continues. A dry line is often the focal point of a supercell. "Fronts are pressure boundaries," he ex-

plains, "but a dry line doesn't necessarily need to be a pressure boundary—you have moist air on one side and dry air on the other side. Usually the western side. Your westerly winds during the day will press dry
air toward the east—so your dry line is advancing east. So right ahead
of the dry line, you'll have southerly winds drawing up warm, moist air.
That sets the stage." He stops for a minute, searching for a more dramatic turn of phrase. The conditions at that time, he concludes, "are
screaming for convection to take place."

Such was the situation on May 3, 1999. Unstable air always has
an enhanced vertical component. Pockets of air that are warmer and
less dense than the surrounding air will rise, cooling to form cumulus
clouds. Often, though, the rising air will reach a warmer layer midway
up the troposphere—a "cap"—that halts its progress.

"The cap is a layer of air that's warmer than the warm air rising,"
says Klee. "If the cap is strong enough, that's as far as the cumulus will
develop." In such a situation, showers may result, but you will see
nothing like the severe thunderstorms that devastated Oklahoma City.
"If the cap is weak, rising clouds will shoot straight through the cap.
And once you're above that, the air is exceptionally unstable."

On the day of the killer tornado, that's exactly what happened.
"The cap wasn't very strong," says Klee. "So the conditions in the middle of the atmosphere were almost perfect to allow convection to go
unchecked." That allowed cumulonimbus clouds to grow into monsters, poking their billowing tops into the stratosphere.

Later in the week, Jacob Klee surveyed the damage near Bridge
Creek, Oklahoma, and he also flew over the town of Moore. "Driving
through the area or flying above the damage gives you the connection
of seeing it with your own eyes. Like, 'Wow, this is real. Here and now.'
It hits a little harder than the TV images. But you're still a bit detached
from what surrounds you. At least I was."

He describes to me how the disaster really hit home only after he
had walked through the debris on foot. "The stench of death was
everywhere. Underfoot and aloft in the remains of the trees were the
little bits that make up a person's life. A bottle of shampoo, a muddied

*Pacific waves hammer
the coast of Southern
California during
El Niño.*
PHOTO BY MARK ROSS-PARENT

*Saw blade embedded
in a telephone pole.
Tornado damage,
Oklahoma, May 1999.*
PHOTO BY JACOB KLEE

Tornado damage, Oklahoma, May 1999.

Photos by Jacob Klee

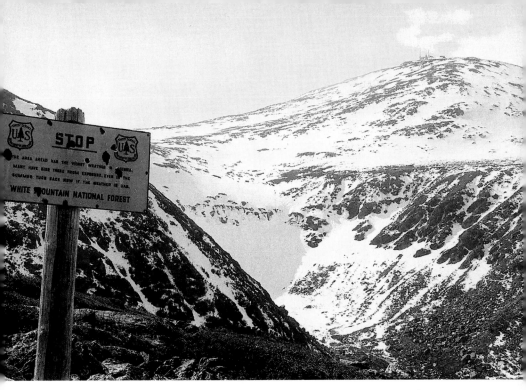

Mount Washington, New Hampshire, "Home of the World's Worst Weather."

PHOTO BY MARK ROSS-PARENT

Glaze ice immobilizes a truck, summit of Mount Washington.

PHOTO COURTESY OF THE MOUNT WASHINGTON OBSERVATORY

*Heavy rime, deposited by freezing fog, immobilizes three probes
on the summit of Mount Washington.*

PHOTO BY ERIC PINDER

Bracing against hurricane-force winds, Mount Washington.

PHOTOS COURTESY OF THE MOUNT WASHINGTON OBSERVATORY

A weather observer de-icing atop the observatory tower after an ice storm deposited more than ten inches of glaze in a single day.

Fueling a plane at the South Pole, summertime.

Launching a weather balloon from the South Pole research station. Helium-filled balloons are launched twice a day in the summer and once a day in the winter. They rise into the stratosphere, carrying instruments that measure temperature, humidity, and pressure. Researchers will track the balloon for 2.5 hours, and can also determine wind speed and direction at different altitudes based on its movements.

PHOTO BY MARK ROSS-PARENT

Abandoned during the long winter night at the South Pole, this "Jamesway" building (or Quonset hut) will slowly be covered by drifting snow. It will be excavated the following summer and used as housing for the influx of additional crew members. (More people work at the Pole in summer than in winter.)

PHOTO BY MARK ROSS-PARENT

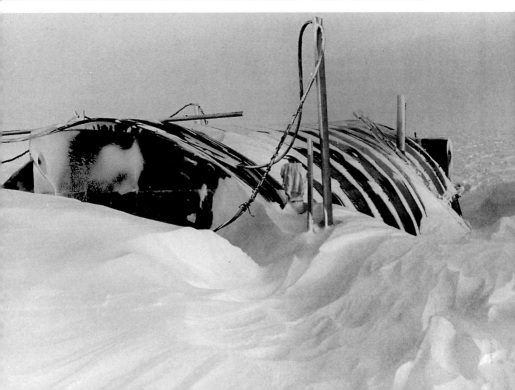

Altocumulus clouds over the South Pole's geodesic dome.

PHOTO BY MARK ROSS-PARENT

A rare visit to the desolate South Pole by a skua. Normally, no wildlife exists at the Pole.

PHOTO BY MARK ROSS-PARENT

Lenticular cloud (middle left of picture), mountains of New Zealand.

teddy bear, a computer disk, a VCR tape fluttering about, a shower pipe partially covered in mud, a baby's shoe, an adult shoe, the frame of a mobile home wrapped around a utility pole, a mangled stereo near a muddy book." It only got worse. "To look and see the grass more or less gone, every tree around reduced to at best a twisted stalk, people milling, looking for anything that might be theirs. This was one of the most saddening things I have ever seen."

I share the sentiment. Wind and weather have always fascinated me because they are beautiful, if sometimes dangerous. And there is a certain satisfaction in overcoming or surviving danger, even if it is done vicariously through the pages of an adventure book or by passively watching a movie like *Twister* on a giant screen at the multiplex. That, I think, accounts for the popularity of the disaster genre. In real life, however, all it takes is a short walk through the wreckage of Oklahoma City to strip away the illusion. It isn't like the movies. The real experience is altogether less pleasant. There is no happy ending.

"IF YOU DON'T like the weather here, wait a minute," is an old Yankee saying, usually attributed to Mark Twain. Probably the phrase originated in springtime, that fickle season of moody skies, gentle winds, and raging storms all sloshed together. Springtime is a half-brewed stew of weather.

Despite the recent thunderstorm, I have few worries about tornadoes here in New England. The river valley where I have lived for eight years traps the air in a topographical bowl and moderates the extremes. Spring snow on the highlands falls as rain in my backyard. A ridge of tree-topped hills shields the village from the blunt thrust of frontal systems. Though tornadoes do occur in all fifty states, no one I speak to recalls a local tornado in recent years. No, to find the extremes I must go elsewhere.

The weather here is fickle nonetheless. In farm country, the air thaws, releasing a pungent odor of old hay, soil, and manure. Each new hour brings additional changes. This morning's air is warmer now, but

I woke to witness a lawn caked with silver frost. My footprints sunk in the stiff grass, which was edged with ice. Yet here and there, newborn flowers poked impatiently through clumps of old, dirty snow.

No one watches the weather with a keener eye than do farmers, who supply the grains and meats that keep civilization going on a full stomach. "You have to make hay when the sun shines," they affirm. In same breath, the whispered truth is this: "Better not bale in the rain."

Spring weather sets the tone for the seasons to come. Ice and snow melt into the soil and the permafrost withdraws deep underground before the plow digs its first furrows in the fields. "Moist April, clear June," the experienced farmer believes. "May damp and cool fills the barns and wine vats." But a late start—or a false start, with fresh buds nipped dead by frost in June—spells disaster for the coming harvest. "A late spring is good for corn, bad for cattle," according to folklore.

Centuries ago, when the majority of the world's population eked out a living in the dust, mud, and rocks of the fields, an ability to read the sky's intent played a major role in survival. The ancients were sky watchers; they bequeathed to us a legacy of weather wisdom to supplement our satellite photos and coiffured meteorologists on the six o'clock news. "If the spring is cold and wet, then the autumn will be cold and dry," advised the ancients. Shakespeare referred to "the uncertain glory of an April day," and people in Wales once said, "April snow stays no longer than water on a trout's back."

In spring, you never know what you'll get. This morning, in a shed behind the barn, young lambs cluster around me in a burst of white wool, eager for their breakfast. I toss a bale of alfalfa in the feeder, then watch as they frisk and play, running haphazardly from one end of the pen to the other. They kick their legs high in the air like children playing soccer. Their *baas* rise in light, fuzzy notes, joining the trills of birds and the low music of the newly thawed breeze.

Spring is here at last, but I feel a certain stiffness to the season. Earth is like a sleeper who has just awoken grumpily and needs to stretch.

SPRING WEATHER SERVES us with a reminder that life and death
never lie far apart. "Better late spring and bear, than early blossom and
bust," I remember reading in a book. "Early thunder, early spring" is
another old expression. But what if the spring season never came at
all? One year it actually happened; winter weather leapfrogged the
vernal equinox in March and persisted into summer. Farmers glared at
the sky and shook their heads in wonder and dismay. The chill of December lingered, refusing to yield.

Whenever a stubborn winter pushes forward on the calendar, it
disrupts the natural order of the seasons. April buds shrivel; half-open
apple blossoms wilt and die in May. In 1816 winter bypassed spring
and summer altogether, smiting hay, corn, and other crops with killing
frosts. People throughout the northern hemisphere, and farmers in
particular, were far from pleased. The season is still remembered as
"Eighteen Hundred and Froze to Death," or "the Year without a Summer."

One early morning in mid-June of 1816, Mr. Chester Collins of
the village of Tyringham discovered a crusty sheen of frost on the
ground and somebody else's horses eating grass in his yard. Neither
was welcome. He jotted down a quick note and delivered it to the office of the local newspaper:

> *Notice: Broke into the enclosure of the subscriber on Saturday*
> *last two 2 year old Horse Colts, light bay, with a star on the*
> *forehead, and white hind feet. The owner is requested to prove*
> *property, pay charges, and take them away.*

That took care of the horses; the frost was not so easy to get rid of.

What was wrong with the weather? Bitter cold spells struck repeatedly in June, July, and August. In the Berkshires of Massachusetts, snow-dappled hills glimmered in the summer sun. Chester
Collins and every other farmer cursed the bad weather, but they never
guessed at its unlikely source.

A year earlier, on April 11, 1815, a two-year-old black stallion colt
owned by Charles Mattoon of Lenox had broken free, or else was

stolen. Mattoon placed an ad in the *Pittsfield Sun,* offering a reward for the colt's return. The disappearance of Mr. Mattoon's colt was unremarkable save for one extraordinary coincidence: at exactly the same time, halfway around the world, a volcano in Indonesia blew itself apart.

THE ERUPTION OF Mount Tambora spewed millions of tons of ash and dust into Earth's atmosphere, dust that intercepted the sun's incoming rays and bounced them back into space. Pumice fell like snow on the ocean, but much of the dust lingered in the high stratosphere and slowly circled the globe, a cloak of darkness. Skies dimmed; the sun turned pale and red.

In 1816, after continuous dry weather and severe frost in May, an essay from Virginia was widely reprinted across the fledgling United States, commenting on the dismal spring.

> *The temperature of the weather with us is very fluctuating—the evenings and mornings generally so cool as to render a fire quite agreeable. The earth is so parched, that the atmosphere is continually impregnated with a fine dust, very injurious to respiration.*

Scant precipitation fell that summer, and often what little there was fell as snow. Farmers scratched their heads in dismay, bent their backs, ploughed under their dead corn, and tried again. And again. And again.

In July, the papers announced that

> *a famine for man and beast seems to stare us in the face. If there should be hay to winter one third of the stock which was kept last season, it will be much more than is expected. . . . The night before last, the frost was sufficiently severe to kill beans, cucumbers, squashes. . . . We hope, notwithstanding, that the God of harvest will not utterly forsake us.*

During a surprise blizzard in June, a man known only as "Winchester" stepped outside his house and waded through snow to build a shelter for his sheep, which were pastured a mile away. He instructed his wife to start searching for him if evening fell before he returned. "June is a bad month to get buried in the snow," he is rumored to have said. "Especially when it gets so near the month of July."

Tendrils of snowy mist clung to his farmhouse like the sticky white gauze of spiders' silk, an icy cocoon. The man kicked open a path with his feet, waved goodbye to his anxious family, and promptly vanished.

Night fell, snow continued to tumble from the sky, and somewhere in the darkness on a slippery patch of ice, Winchester fell. Perhaps he struggled to his feet, only to slip and fall again farther down the road. No one knows. Three days later, a neighbor stumbled across his stiff, frozen corpse, buried by the drifting snow.

HISTORIANS MAY WINCE at the Winchester story, dismissing it as a preposterous tall tale. Buried by a blizzard in June? Whoever heard of such a thing? But I've learned to expect the worst—and best—from weather, especially in springtime. Not even a blizzard could surprise me. After all, on my birthday in late May a low pressure system once swooped across from the Great Lakes and buried the yellow dandelions under twelve inches of snow. The fresh green lawn lay wrapped in a white sheet all afternoon, a birthday present slowly opened by the thawing air.

Springtime has arrived on schedule this year, but it is fickle. Tree branches grow mitts of leaves that wave and sway in the wind. The first hard knock of a woodpecker's beak on bark acts as an alarm clock for the bears who have snoozed lightly all winter long.

In a far-off grassy field that ripples and dips over the hills, I stride along to collect a flock of forty sheep and move them to better grazing ground. A tiny border collie jogs at my side, tongue awag with joy. Every so often she stops to sniff a woodchuck's hole or a flower or a stone; then she sprints to catch up.

A bucket of feed swings in rhythm against my jeans. With this supply of grain in hand I open gate after gate, finally prepared to lead the flock of ewes to the next field. Fresh grass awaits them.

The sheep spot me, or perhaps they see only the bucket of food in my hand; they *baa* in unison. The ewes cluster together, a sea of wool. Overhead, a crow caws. The wind whispers, still moist from the early-morning storm.

"When sheep or oxen cluster together as if seeking shelter, expect a storm" is the advice of weather folklore. I glance up at the sky. But the collie barks a warning—a crisp jolt of noise. I look back. One ewe stands alone, far behind the rest of the flock, her wool matted and dirty. A creek of muddy water gurgles at her feet in a gully through the meadow where the melting snow slowly trickles away. That stream will soon dry up; it won't survive the heat of May.

As soon as the other sheep are safely locked away, contentedly chewing the tall wet grass in the next field, I hurry back to investigate the solitary ewe.

She ignores me. When I walk closer, she refuses to give ground. The rest of the flock is now far out of sight, but this lone ewe does not call out or rush to join them. Sheep are herd animals; their behavior is etched deeply in their genes. Instinct commands them, "Do not stand alone." For the ewe to behave in such a strange, aloof way, something must be very wrong.

The rims of the sheep's eyes look black, crusty, and unhappy. Wet wool mats her back. I swing open the latch that separates us and hear a thump of metal on wood as the latch falls back into place.

For the first time, the ewe shies away. I hold out a tempting hand-ful of grain, but she fails to take the bait. She backs away.

I step forward—and stop. Inches from the toe of my boot lies the tiny carcass of a lamb. Another small carcass sprawls beside it, soft and white, scarcely bigger than my hand. The ewe bobs and weaves her head. She issues a weak *baa* like distant wind. Now I understand.

Lambing finished weeks ago; the most recent crop of young lambs had huddled around me earlier this morning. All of them were agile, alive, tall as my knee. But each of the two lifeless husks lying still at my feet could easily be held in the palm of a hand. They are buds that barely sprouted, leaves that never grew.

Surely they died weeks ago, and were then ripped open and gutted by fox or coyote and abandoned in a clump of grass and mud. Their wool is still soft white, almost shiny, but the lambs themselves look like puppets, empty sheaths of wool with gruesome bony puppet heads. Their bodies are empty. Their blackened eye sockets crawl with flies. Possibly those eyes never opened, never saw the spring rain or the sun on the grass.

I now believe the lambs were early arrivals, born before anyone noticed, before the flock of still-pregnant ewes was brought into the barn to give birth. Born before anyone had reason to suspect they were alive. Any flower that blooms in a too-early spring will perish, crushed in the grip of a returning winter.

I wonder, does the ewe still remember them, after so many weeks? How long is a month in the life of a sheep? She must have been herded away from this field weeks ago with the rest of the flock, only recently to return with the warmth of the sun. Yet still she guards her lambs, sensing something vital missing inside those empty shells of wool, those half-opened white blossoms that died in the midst of spring.

12

The Invasion of Summer

WHO WOULD HAVE GUESSED THAT ONE OF THE WISEST QUES-tions ever asked about wind and weather issued from a talkative but fictional teddy bear named Winnie-the-Pooh?

In the classic children's story by A. A. Milne, Pooh ponders aloud,

Isn't it funny that a bee likes honey?
Buzz buzz buzz, I wonder why he does.

Surprisingly enough, the long-winded answer to that rhyme is detailed in countless books, scientific journals, and argumentative letters-to-the-editor in newspapers around the world. Despite Pooh's reputation as "a bear of very little brain," the riddle he proposes cuts deeply into the past; it requires a probing look at a revolution in science sparked by the writings of Darwin and Wallace. In just a few words, Pooh hints at a relationship between living creatures and the wind extending across a vast span of hundreds of millions of years.

It will take some time to explain.

YEARS AGO ON a summer morning while still living in Massachu-setts, I awoke at dawn with that old, half-forgotten riddle from Win-nie-the-Pooh inexplicably echoing in my head. Soon I discovered why. A honeybee 12 inches from my ear buzzed angrily, emitting a high-pitched whine that needled at the raw edge of my semiconscious brain. It woke me like an alarm clock. Bleary-eyed, sprawled on one side with my left arm tingling from a lack of circulation, I focused first on a calendar tacked to the wall. One day was circled: June 21, the first day of summer. Then I rolled over to face the bee.

The insect darted forward with an angry flutter of tiny wings, its thorax and stinger an inch from my nose. The thick glass of my bedroom window separated us, so I resisted the urge to cower and dodge. To the bee's eyes, that window appeared as nothing but a puzzlingly dense slice of air. No wonder it was angry. The bee skated across the glass surface, balanced on the tip of its stinger. The sun rose behind it and cast a menacing, alien shadow on the far wall. A tightly wound coil of black and yellow stripes decorated the bee's body. Like a briar thorn come to life, it flew repeatedly at the glass pane; its delicate wings were a blur of gauze.

Popular belief maintains that a bee's wings never get wet, possibly because bees are thought to take refuge in their hives as humidity increases before a storm. Wings laden with moisture force bees to fly low to the ground and close to home. A rhyme of folklore turns this casual observation of insect behavior into a general rule of thumb:

When bees to distance wing their flight
Days are warm and skies are bright;
But when their flight ends near their home,
Stormy weather is sure to come.

On that long-ago summer day, the bee on the windowpane seemed in no hurry to leave, so maybe I should have anticipated fair-weather skies. The evidence supported that assumption. Sunlight poured down like bright liquid through the leaves, soaking the ground with warm vapors. Sharp, dark shadows flickered and winked on the grass. In the trees, I saw the palms of maple leaves snatch at the wind so playfully that I almost wanted to believe the ancient, mistaken assumption that wind is set in motion by the swaying of trees, rather than the more logical alternative. The Greek philosophers' errors have always possessed a poetic appeal that lends itself to a summer day. Hurricanes, Aristotle believed, are caused by "evil winds falling on good winds, with resulting moral conflict." But that opinion, right or wrong, did not concern me at the time.

I pulled shut the inner window, so that the bee—it had navigated through an obvious hole in the screen—lay trapped between glass and mesh, caught like a paramecium between a microscope slide and a cover slip. The invader buzzed, outraged. I studied its behavior. For half an hour, that bee wandered purposefully across the glass and tested for weaknesses. The friction of its active wings created the buzzing noise; I listened all through breakfast. I had no choice. The first invader of summer had arrived.

THE HOT, MUGGY "dog days" of summer owe their name to the ancient Egyptians, who watched the bright star Sirius—the Dog Star, in the aptly named constellation Canis Major—rise in the east near daybreak every morning from early July to mid-August. They believed that it contributed warmth to the blistering summer weather. (The flip side of a "dog day" is a "three-dog night." Back in an era before central heating, when the only available "electric blankets" were dogs and cats sleeping atop their masters' beds, a particularly cold night might require three dogs.)

The solstice—June 21 or 22, depending on the year—marks the onset of summer, and ushers in a season where the gripping cold claws of winter retract till November. Snowfall fades away like a distant memory; flowers bloom. The apple tree in my backyard tosses white petals like bouquets across the lawn.

Contrary to popular belief, the earliest sunrise and latest sunset

*Likewise, it is often assumed that during the equinoxes in March and September, day and night are exactly equal at 12 hours apiece. That's incorrect—though for an entirely different reason. Sunrise and sunset times are measured from when the center of the sun's disk reaches the horizon. Since the sun is a disk, not a point, that means a sliver of the sun will be above the horizon and casting light a few min-

do not necessarily occur on the solstice.* Astrophysicists are among those who understand why, and the full, technical explanation would almost require a chapter of its own here. The tilt of Earth's axis and the eccentricity of its elliptical orbit around the sun both play both roles. Suffice it to say that the "solar day" is slightly longer than 24 hours at the time of the solstice. (It's slightly shorter than 24 hours at the time of the equinox.) "Solar noon" is the time at which the sun reaches its highest point in the local sky, and a solar day is the period between one solar noon and the next. A solar day and a "clock day" differ: according to our clocks, every day is exactly 24 hours, no matter what time of year. This discrepency between clock time and solar time is the answer. Clock time is always catching up to or passing solar time. For that reason, the latest sunset will occur a week or two after the summer solstice, while the earliest sunrise may happen a week or two in advance in temperate latitudes. Confusing? Throw away your quartz watch and use a sundial instead, and the whole problem goes away.

In any case, the summer solstice is always a precipice on our annual calendar, a point in time beyond which nights lengthen as the warm season plunges toward autumn.

On June 21, the sun swings in a high arc across the northern sky. Along the Tropic of Cancer—an imaginary line circling Earth at 23.5 degrees north latitude, about 80 miles south of Key West, Florida— the sun hangs in the zenith at noon. Its rays are short, penetrating, and hot. North of that line, the sun is never visible directly overhead. It always gleams at an angle, so that the closer you go to the pole, the lower the sun arcs across the sky. Regional climates quickly turn cold and shivery as the sun's rays weaken.

The irony of summer, for people who live in the northern hemisphere, is that Earth now orbits farthest from the life-giving furnace of our nearest star. The planet reaches aphelion, the most distant point

---

utes before "sunrise" technically occurs. Days during which there are 12 hours of light and 12 hours of darkness actually occur about 48 hours on either side of the equinoxes.

in orbit, in early July. The elliptical revolution of Earth stretches around the daystar like a rubber band pulled loosely between two fingers—it forms a squashed circle, imperfectly round. In summer we are farthest from the fire. In icy winter, no matter how much our logic rebels against the idea, we dip a little closer to the flames.

Incoming solar radiation is actually about six percent greater in winter than in summer, but that's not enough to make much difference. Earth removes itself to a distance of 94.5 million miles from the sun by July 6, compared to a closer approach of 91.4 million miles on January 2. But a few million miles is only a short hop on the astronomical scale and has little effect on everyday weather. Instead, the tilt of Earth's axis determines the seasons. Solar rays can either pass through the atmosphere straight-on, as happens in the tropics, or at a long, sloping angle, such as occurs at the poles. Air molecules in constant motion get in the way. Clouds bounce rays back into space, and billions of air molecules absorb the energy of many sunbeams before they can strike the planet. The greater the distance that sunlight must pass through air, the more anemic the sun's rays feel by the time they end their eight-minute journey across the solar system and impact Earth's surface.

Because of the tilt of the axis, the northern hemisphere leans toward the sun in June, making the sun appear ever higher in the sky at noon. The rays grow more intense, and the grass flushes a thick, lively green. The South Pole turns bashful at the same time and wobbles away for six months of deadly cold night. At the bottom of the world, the unseen sun swings in circles below the horizon until the long-awaited break of day.

THOUGH THE LONGEST day occurs in June, we experience our hottest average temperatures in July and August. Humidity rises; the heated air is lighter, less satisfying to the lungs. Even though nights are growing longer, soil and water still heat up more hours per day than they cool after dusk. That does not change until September. An annual heat lag carries warm temperatures well past the solstice to the start of autumn. Perhaps a celestial accountant of some sort adds up all the hours and decides there is a surplus of sun; we are due a refund of warmth, at least until winter's chill balances the account.

In contrast to summer, winter to me is a simpler, slower season, a time when warm-blooded mammals burrow deep underground or else sleep under blankets of twigs and dirt until more amicable weather thaws the air. Many birds flee south. Insects pupate, immobilized by bitter cold, or shrivel up and die. Winter is a lonely, contemplative time. Summer, by contrast, demands our attention. It is always a loud and crowded time of year.

Who invites the hubbub of summer? Somehow it sneaks up unexpectedly. This morning in my kitchen, on the bottom of a drinking glass, a thin residue of orange juice bubbles. The liquid is inhabited by at least fifty ants; they have jumped in and learned how to swim. The welcome frost of a winter or autumn day would quickly grind their activities to a halt—but here they are. Their knobby black bodies wade in the liquid, or else lie still on one side, like beachgoing tourists bloated on sun and drink. Where have they come from? What sign, trail, or scent has led them straight to my breakfast juice?

Outside, June's clear skies send sunbeams skittering across the grass and streaming through the windows. I dimly recall that I had left my glass there, half-empty, early in the morning. After I'd taken a single sip, the phone rang and business intervened. I set the glass down and forgot it. But the ants did not let the juice go to waste. Did a single, aimlessly wandering scout locate this mecca of liquid sugar and report back to the rest? Is it possible that insects can smell a half-inch of citrus juice in the bottom of a glass from some distant hole in a corner of my backyard?

Any scent, fair or foul—flowers, mints, fresh-baked bread, even the musky odor of wet dog—consists of tiny traces of whatever original substance it comes from, carried invisibly in the circulating air. Wind never rests. It peels a microscopic sheet of atoms off the surface of my orange juice and disperses it on the breeze.

———

A BAG OF blueberry muffins, sufficiently wrapped so that I can't smell them, pops and crackles mysteriously. I open the bag to discover yet another ant colony at work. So much for lunch. Light pours into the paper receptacle, and the ants scurry.

Ridding the house of this sudden infestation proves to be an all-day affair. A cluster of ants colonizes the left-hand corner of my desk. Perhaps they are feasting on tiny crumbs from some improperly cleaned-up snack. Why else would they stay?

With the edge of a sheet of paper I scrape the ants off the wooden surface into a bag. They scuttle madly as my hand touches the wooden surface; I watch them flee. Often they run ahead of my sweeping palm and beat me to the task, leaping in desperation off the edge of the desk and landing with a plop in the open paper bag below. Their comic flight brings to mind a badly made horror film, with hundreds of screaming extras stampeding down the streets in advance of the monster's footfalls. They sprint in mindless terror, the way a town full of human beings would surely scatter if a giant hand reached down suddenly from the sky and started squashing people at random, or else stuffing them into a bag to be carried off and disposed of later in some odious manner.

Once I have trapped the entire colony, I curl up the bag and toss it into a garbage can. But later that night, I wake with a start; I listen. A rustling noise emanates from a hidden corner of the room—the scraping sound of hundreds of tiny feet crawling on paper. The bag pulsates as if it were alive. I think of Poe's tell-tale heart, beating warm pulses of guilt under the floorboards. Dozens of insects now patter across the inner surface of that sealed paper bag, around and around, finally coming back again to their starting point. There is no escape.

I repent and let them outside to find their way. Sleep finally takes hold. The summer wind settles into silence.

IN ADDITION TO detecting my lunch from some faraway corner, insects and other animals can gauge the mercurial changes in the atmosphere with greater accuracy than can human eyes and ears. Ants are mobile barometers, indicators of bumps and bounces in atmospheric pressure. If you ever see ants run wildly, and you are not deliberately provoking them with ant spray or other horrors, then a fair summer day is probably in place; low pressure and thunderstorms are unlikely to ensue. "Expect stormy weather when ants travel in lines and fair weather when they scatter," the folklorists tell us.

Some insects, such as ants, release pheromones to produce a scented trail; it allows them to follow one another. In the case of high pressure and fair weather, the air is sinking through the atmosphere, keeping the scented trail close to the ground. As a result, the ants seldom get lost. But when low pressure creates stormy weather, air rises and the pheromone trail lifts up beyond the reach of wingless ants. To get where they are going, they must follow each other with great attention, giving rise to the old adage about ants marching in a line before a storm.

Just as dogs and cats are said to act skittish in advance of earthquakes, like living seismographs, we can use observations of other animals for hints of the coming weather. Folklore is especially rich with the wisdom of creepy-crawlies:

Ants that move their eggs and climb,
Rain is coming anytime.

"When spiders work at their webs in the morning, expect a fair day," another weather book informs me. The assumption, apparently, is that spiders do not waste time with elaborate weaving if wind and rain will soon rip apart their artistry.

Snakes are thought to hunt for food before a rain, and thus to be hard to find immediately afterward. "Snakes and snake trails are often seen near houses and roads before a rain," one book of folklore instructs. Low pressure in advance of a rainstorm brings out the snakes—and countless other creatures—in greater numbers, so they are temporarily easier to see. And then the rain pours down.

———

HOW LONG, IF I sit still, will it take the maw of summer to devour me? All these ants, bees, moths, and spiders—not to mention fungi and molds—must constantly creep, slither, crawl, and glide, animated by the warmth of air.

Without wind, without the vapors sent swirling across a newborn planet 4.5 billion years ago, none of these complex life-forms could stir today. The orange juice sitting on my desk would not exist. The deciduous fruit tree from which it came would never sprout in the soil. I would not be here to drink it, in any event. Without our ocean of air, life would never have left the sea—if it ever would have formed at all.

In summer, bees flirt with nectar in the flowers. Seeds germinate; thistles grow. Wind hoists into flight the white seeds of dying dandelions. It also catches and carries hard, wooden seeds as they tumble from the trees. I watch a maple seed ride the breeze and twirl like a helicopter rotor from a high branch. Slowly, wind pushes all of these packaged nuclei of life across the continents and encourages them to grow.

Life only gradually learned how to harness weather to advance its own needs. Deep in the Precambrian era, long before the age of dinosaurs, blue-green algae and photosynthetic bacteria digested the poisonous fumes of the early atmosphere and pumped oxygen into the sky. Two billion years ago, enough oxygen had accumulated to form an early ozone layer, shielding the land from the sun's harmful ultraviolet rays. Safe at last, life emerged from the ocean and took root on the naked land.

Deep in the Silurian period the first fern spores drifted on wind and water, colonizing the banks of streams and eventually expanding across the barren rocks of unclothed continents. The first breeze was a gardener—a very patient one, planting and harvesting for millions of years. Eons before the first human sailors raised a sheet to the ocean gales, an early plant—long since extinct—cast a spore to the winds just to see where it would go.

Wind rapidly carried life across the globe, but the colorful diversity of life so familiar to us today took a long while to arrive.

Geologists tell us that the first spring in the poetic sense occurred in the Cretaceous or late Jurassic era, more than 140 million years ago, when primitive flowers blossomed for the first time. Much later, in the rich summer of the Cenozoic era, ever newer and more colorful species of flowers opened their petals above the gravestones of dinosaurs; mammals crawled out of the shadows unchallenged.

Primitive flowering plants probably appeared some time during the Jurassic dominance of the dinosaurs. The first flowers were simple, little more than colored leaves draped around a seed. But they quickly exploded with diversity, flushing the dull green-and-brown world with unexpected delights. By the time of the cataclysm that marked the end of the Mesozoic era, 65 million years ago, the world had changed forever. Whatever tragedy triggered the abrupt extinction of more than half the species alive—whether a comet or asteroid exploded against the coast of Mexico, or the global climate changed of its own cyclical accord, or both—it truly marked the passing of an epoch. At the start of the Mesozoic era, 245 million years ago, the world was simple and green. No leaves turned color with fall. Once upon a time, as the storytellers say, blossoms never bloomed.

By the end of the Mesozoic era, flowering plants of all shapes and hues had evolved and learned to harvest the wind, growing aerodynamic seed casings like the winged maple seeds of today. Similar seeds with fins or wings drifted ever farther on the breeze, carrying their genetic material to new corners of the globe and ensuring the survival of the parent plants' DNA.

To better reap the wind, plant life in due time coaxed specific members of the animal kingdom to help carry them across the sky. Think how many grape seeds from California get shipped to supermarkets around the world and are eventually spat into distant soils; the process hasn't changed much in 140 million years. Long before the age of refrigerated trucks and airplanes, insects served the same purpose. Mosquitoes and bees evolved wings of their own in order to navigate the swirling currents of the atmosphere; they carried pollen from place to place. Plant life learned to bribe them with scents, honeys, and colorful dyes. It is a folk belief even today that certain bees grow attracted to the colorful shirts of people outdoors, mistaking them for flowers.

The first bees on Earth flew hungrily from one source of nectar to the next, brushing against each flower with small, furry bodies specifically designed to wallow in the fragrant, life-bearing dust at the center. Coated with pollen, they transported the stuff of life to distant sources of fertilization, and thus helped the plant life thrive. A symbiosis emerged.

Soon after the first flowers had bloomed so suddenly in the deep past, the advent and refinement of bees and similar creatures was inevitable. Today, 700 species of butterfly inhabit North America, fulfilling the same ecological function of feeding on and transporting fertile nectar. The names of butterflies, and the decorative patterns of their wings, brilliantly echo the prisms of flowers on which they depend: swallowtails, coppers, blues, and American painted ladies, to name just a few.

Colors did not exist in such abundance in the monochromatic world of the distant past. The first red berries finally appeared in a warm geologic summer at the end of a dying green era, deep in the Mesozoic. Earthly breezes carried the seeds, and in doing so determined the biology of hummingbirds, butterflies, and bees. Wind is what inspired a true ripening of life on Earth.

13

Spirit of the Winds

A satellite image of Hurricane Fran scrolls repeatedly across my computer screen. The image depicts the blue expanse of the Atlantic Ocean pressed firmly against a ridge of brown and green pixels representing North and South America. In the upper corner, the hurricane itself practically jumps off the screen, but its attack is silent and swift. It is a circular blotch of white light stamped like a bull's-eye just off the coast of North Carolina. The tops of the clouds shoot ten miles straight up to the limits of the troposphere. I feverishly imagine a cloudy pillar of convective energy billowing out of the monitor toward my eyes. The curved edges of the storm are equally violent, curled in tight bands around the center. Nowhere else in the Atlantic do the clouds contort and twist themselves in quite this way.

The storm slams into Cape Fear just after 8 P.M. on September 5, but I hear no wind, no screams of panicked beachgoers (the foolish few who have failed to evacuate by now), no surge of the tide. Wind spins and howls counterclockwise around the eye; gusts of 120 mph uproot telephone poles, and bulldoze through large houses as if they were fragile models made of balsa. Waves claw at the beaches, trying to drag land back into the sea. A storm surge—the weight of the ocean's surface literally sucked into the sky by low atmospheric pressure, then splashed onto shore by the wind—floods the coast and carries away the flotsam.

I comfortably watch the destruction from my office at the weather observatory a thousand miles away, perfectly safe. A wall of seawater ten feet high rumbles through the once-dry village of North Topsail Beach and knocks the town hall off its foundation. Low-lying roads vanish beneath a muddy sea. A snarling breeze saturates the air with water and wind, but no spray hits my eyes. This hurricane consists of a billion tons of moisture packaged into a wheel of wet wind spinning across the ocean until it collides forcibly with land; even so, I feel no fear. Calmly, I peck away at the computer keyboard and call up the latest damage reports. Modern technology has reduced the distant storm to a harmless pattern of colored dots.

The satellite image I see is a snapshot taken by an orbiting camera 22,500 miles above Earth. The picture changes every second as a time-lapse re-creation rolls the eye from the ocean to its farthest encroachment on land. The spiral limbs flail, like a child's arms in the midst of a tantrum. They swirl around an ominous but calmer focal point, the eye, which stares out of my computer screen, unblinking and pernicious. No wonder sailors in past centuries imagined Cyclopean monsters among the terrors of the sea. Hurricane Fran's single eye glares up at the heavens, even as its massive aqueous body lumbers ashore.

HURRICANES ROUTINELY DEVASTATE the coastline of the United States in the summer and fall seasons of each year, causing billions of dollars in damage. Hurricane Fran's deluge has already run up a bill topping three billion dollars. In 1989, Hurricane Hugo, a similar-sized monster from the tropics, flattened nearly 9,000 square miles of forest and killed 49 people in the Carolina region, with damages estimated at eight billion dollars.

Technically, a hurricane-force breeze is any wind 74 mph or greater. Down in the stormy Caribbean, winds of this magnitude rip roofs from houses on a regular basis; they juggle sailing ships like toys and then cast them in heaps of broken wood onto the beaches. When a hurricane hits land, whips of lightning often crack across the sky in the region of the outer spiral bands; tornadoes may plummet from the bulging roof of clouds to ravage the ground. Heavy rains pour down like waterfalls. Winds shriek a warning that few alive are left to hear.

At sea, hurricanes can kick up waves more than fifty feet high, taller than ships' masts. Heat from the warm water of the tropical ocean pumps inside these storms and fuels their chopping arms.

In 1900, the Galveston hurricane lashed against the coast of Texas, killing more than six thousand people with a sudden burst of water and wind. The ocean spread wide its jaws and licked the land with a watery tongue. "It ripped the coast right off!" exclaimed one excited historian, describing the scene from a safe vantage point 98 years later.

The New England hurricane of 1938 exhaled a barrage of wind and ocean spray off the Connecticut shoreline. The storm's breath spit clouds and rain deep into Massachusetts and Vermont, uprooting millions of tall trees like so many weeds. "I remember when I was eight or nine years old," an elderly woman named Florence once told me, describing the devastation of her hometown in northern New Hampshire, where she has lived all her life. "Our neighbors had a pear tree and the wind tore it out of the ground. I hope we never have another hurricane like that one."

Perhaps it's no surprise that certain people in the hurricane-prone Virgin Islands piously celebrate a special holiday—Hurricane Supplication Day—at the end of each July. No one goes to work. Instead, they pray for protection from storms from the sea. But the frantic residents of Eastern Seaboard cities share no similar holiday. No one in the United States cheers or celebrates the arrival of Hurricane Fran this morning. Boarded-up windows buckle and brace in the breeze. Ghost towns, abandoned by their tenants until the worst is over, flood with salt water and rain. On the fringes of the storm, terrified families huddle in basements and hope the wind will hush. Shingles fly off the sides of buildings and knife through the air; no one goes outside.

Fran finally weakens and starves as she slides her massive body over land, but for a long, weary day, the gusts still top hurricane force. One day later this wheel of wind rolls north into Virginia and gets officially downgraded from a hurricane to a tropical depression; sustained gale-force winds continue at a mere 40 mph or greater. As if to protest this indignity on the part of the National Weather Service, Fran throws a handful of short-lived tornadoes against the ground, sending people running in terror. The rising slope of the Appalachian Mountains intercepts the sleeves of the storm, still laden with tropical moisture, and wrings them dry. More than a foot of rain cascades out of the sky from the Carolinas to Pennsylvania. Before the worst is over, 34 people die.

More tornadoes spike downward from the bulbous mass of cloud. The whipping gyres channel the strongest winds ever known into relatively small areas; a whirlwind of this kind once slammed a wooden two-by-four through a car window and left only a clean hole—the glass did not shatter. Tornadoes always form beneath cumulonimbus clouds—thunderstorms. And what is a hurricane like Fran but a cluster of thunderheads clumped together, fueled by the angry heat of the tropics, wreaking havoc upon the land?

In complete security, I watch the remains of Hurricane Fran push inland. The satellite image and accompanying analysis dissect the anatomy of this beast in great detail—isobars, precipitation totals, windspeeds at different points in time—and yet I experience no wet slap of rain on my face, no shove of tornadic wind against my back, no loss of property or life, no fretting or fear. Compared to the unfortunate few who encounter Fran's wrath firsthand, I know nothing but numbers. The power of a hurricane cannot be quantified so easily.

Another day passes, and the remnants of Hurricane Fran finally dissolve into a harmless breeze in Pennsylvania. A storm that thrived for almost two full weeks at sea dies swiftly once it runs aground. All hurricanes behave this way. After a cyclone hits land, it lacks a plentiful supply of water vapor to inhale as food. Soon it starves on the dry ground and extinguishes itself with a sigh.

THE WORD WE use today to describe these powerful tropical storms originated many centuries ago when natives of the Caribbean dubbed the vengeful god of weather *Hurakan*. This spirit of West Indian winds also wreaked havoc on the fragile wooden ships of the first European explorers, who sailed in search of spices and gold. When storms at sea sank their ships, the Europeans who survived called them hurricanes.

Fleets from Portugal, Spain, Holland, England, and France set out to conquer the New World in increasing numbers. Wind filled their sails and pushed them across the sea on the easterly trade winds that straddle the globe in the tropics. Although these explorers brought along the baggage of their own superstitions and customs, and therefore scoffed at any local warnings about the pagan god of storms, they still respected the power of wind on water. The sea breeze miraculously tapped into the energy of the sun and provided power for their ships in an otherwise primitive world. Wind was savior and destroyer both, depending on its mood. Ships' captains learned that the breeze could either carry them to wealth in the "Indies," or sink them to the watery grave of Davey Jones' locker.

Tropical air surged into the sails of Columbus' ships and powered them across the Atlantic, but the breezes he met were mostly kind. Columbus never encountered a hurricane on his first, famous voyage. In 1495, however, the colony he founded on Hispaniola was destroyed by a storm from the sea.

In 1609, a hurricane attacked a fleet of English settlers and dashed several of their ships against the coast of Bermuda. A survivor's harrowing account of this storm trickled back to England and influenced William Shakespeare as he wrote his final play, *The Tempest*. The spirit Ariel took credit for the meteorological turmoil that shipwrecked Miranda's husband-to-be; he even tossed in a pyrotechnic display of St. Elmo's fire to boot: "I flamed amazement," Ariel boasts in Act I.

The true story was equally gripping. "It could not be said to rain, the waters like whole rivers did flood in the air," wrote colonist Sir Thomas Gates, who later settled safely in Virginia. He began his tale with these words: "A dreadful storm and hideous began to blow out of the northeast, which . . . at length did beat all light from heaven." He goes on fearfully: "Once, so great a sea brake upon the poop . . . as it covered our ship from stem to stern like a garment or a vast cloud."

SIR THOMAS WAS not the first sailor nearly capsized by a storm at sea. Witnessing the advance of a hurricane, the militaristically minded captains of Old World vessels saw an indefatigable armada of clouds. Mightiest of all storms, hurricanes were much feared but little understood.

Greek antiquity tells the tale of the explorer Odysseus, who was once given a strange gift by Aeolus during his ten-year detour on the way home from Troy. Aeolus served the Olympian gods as Keeper of Winds; to aid his famous sailor friend, he handed Odysseus a leather sack, tightly bound, which held all the storm winds. No danger could hit so long as Odysseus kept that sack safe.

Of course, Odysseus' sailors assumed the sack contained gold and treasure, which they craved; they greedily pried open the drawstring and freed the raging winds. The gusts pushed Odysseus far off course, all the way back to the island of Aeolus, who angrily withdrew his favor. Later, the savage winds struck again. This time, torn by the maelstrom, the sails shredded and the ship went down.

Odysseus inhaled deeply, hurled himself into the water, and swam for his life. He swallowed the brine and wallowed in the sea for hours—possibly days—until at last the waves cast him dripping onto the sands of an island, where a new adventure awaited. He crawled onto land in rags, having been triumphant long ago at Troy but all too easily humbled by the power of wind.

Scandinavia's mythology also holds a rich trove of tales about ocean maelstroms. But the ancient Vikings, famous for their sailing prowess, seldom if ever encountered hurricanes. They sailed mostly in the boreal North Atlantic, and hurricanes originate only in the tropics or wherever large bodies of water acquire surface temperatures of 80 degrees Fahrenheit or higher. (Although the ships of the Vikings no doubt ran afoul of the cold breath of extratropical storms, some of which heave waves to heights of 100 feet as they travel near the Grand Banks and south of Greenland.)

Hurricanes seldom drift far north in continental Europe, though a tropical storm once hit the coast of Great Britain and flattened many trees. Out in the western Pacific Ocean, hurricanes—called typhoons—also rage across the water. But so wide and vast is the Pacific that these storms rarely encounter land. They weaken and disappear unnoticed, unheard, seen only by a handful of unhappy sailors and the distant, unblinking eyes of weather satellites orbiting the globe.

ALL MY LIFE I have successfully dodged hurricanes' blows. One time I witnessed the dying spasms of a hurricane run ashore like a beached whale; it plowed across the water onto land and quickly ground to a halt. The genesis of hurricanes occurs far, far south from the land where I live. But sometimes a storm will spiral out of the tropics and rub angrily against the coast of New England.

The radio tonight—I'm listening to a station out of Boston—blares warnings of yet another hurricane that threatens the eastern seaboard. It has been a busy year for hurricanes; in the alphabetical system for naming tropical storms, we are already up to the letter O. World War II meteorologists took to naming tropical storms after their wives and girlfriends, but in 1978 and 1979 the National Weather Service started to alternate male and female names. The cycle of names repeats every six years, except in the case of exceptionally violent and destructive hurricanes like Andrew and Camille, whose names are never used again—a dubious distinction.

It is now early October, the heart of the hurricane season that stretches from June through November. As I listen to the radio report, the atmosphere over North Carolina is already aswirl with clouds and wind. The newscaster's voice shakes with excitement—or anxiety. He stumbles over the name of a town in Florida, where 80-mph gusts are currently making raindrops feel like nails driven against the skin of anyone reckless enough to walk outside. The man corrects himself twice, then clears his throat on the air.

Speculation that the storm may cling to the coast and move up into Long Island and eventually New England is rampant. A trough of low pressure digs down through the middle of the whole continent like a capital **V**, with the point protruding through Mississippi and Alabama into the warm waters of the Gulf of Mexico. Hurricane Opal balances on its tip. The storm is expected to slide northeast up the side of the **V**, essentially following the path of the jet stream. Current computer models project an arrival in New England over the next few days. In the meantime, Opal pounds away at Florida and advances slowly to the north. How quickly will it weaken once it leaves the water? That is the question on everyone's mind.

A day passes, then two. A worried man from Boston calls me at the observatory to ask, "Will I have a house tomorrow morning?" But the pattern has already changed, and the bulk of the storm's wrath is vented to our south and west. The trough steers the eye of the storm elsewhere.

Just after midnight, a TV weatherman in New York jabs his finger at the latest picture of Hurricane Opal, a splash of white clouds washing over Buffalo. The storm has long since run aground, but strong winds and moisture refuse to die. The meteorologist points to a computer-generated recap of its progress; I see a giant wave of wind and clouds roll up the East Coast and crumble across the hills of Pennsylvania, sending a spray of rain into New York and New England. The image is replayed on the screen again and again, while the weatherman stands in front and smiles proudly, as if he himself has just surfed up on a rain cloud from the coast of Florida to tell us the news.

Since the remnants of the hurricane hold a package of warm air from the tropics, temperatures in the northeast are on the rise. Wind, curling counterclockwise around the depression, flows out of the southeast. "Today it will hit eighty degrees in New York!" beams the weatherman. The news appears to make him weak with happy disbelief; his voice lingers over the words "eighty degrees," as if reluctant to let them go. He smiles again and gives the audience an excited little hop.

"A HURRICANE IS like a Maine divorce," quips a burly New England fisherman who doubles as a park ranger on the mountain. "Either way, someone's going to lose a trailer."

Hurricanes pack a punch unequaled in sheer power by any other weather event. Even a tornado, with stronger average winds, is puny in terms of overall size. The slap of a quick summer thunderstorm is easy enough to shrug off; the wind does nothing worse than comb knots into people's hair and snap the twigs of trees. A sudden gust will perhaps make you stumble, but it cannot knock you down. A hurricane, though—that's a different story.

I once watched the wind skip across the ocean shaping waves the size of elephants; they stampeded onto a beach in Maine and hurled themselves against stony crags. The ocean inhaled and exhaled deeply, an undercurrent to the constant hiss of the breeze. But even then, the center of the hurricane lay 200 miles out to sea, slowly suffocating in the icy water of the Atlantic. The beach where I stood received only a glancing blow.

Even without the regular impact of hurricanes, mighty winds howl in the North Country where I live. The breeze often bites with bitterly cold teeth. Air masses that hover over New England are often separated from the warmer air of the Southeast by the invisible line of the jet stream. Winds of comparable speed to hurricanes routinely convulse around the White Mountains, so the equivalent ferocity is no stranger here. Standing in such a tempest is the closest I have ever come to what the unlucky inhabitants of Florida must experience each time a hurricane shreds their roofs and redecorates the interiors of their houses up and down the beach.

I've struggled to stand amid the onslaught of hurricane-force winds as they pummeled the weather observatory on Mount Washington. In a breeze that topped 100 mph and never paused for breath, two fellow crew members and I ventured foolishly outside to challenge its strength. As the observer-on-duty, I did not have much time to frolic with this particular tempest and soon returned indoors. Later, one of my colleagues told me in an awed voice how the third man was lucky to have crawled back in one piece. "He knew the wind would knock him flat in a second, so he rolled out onto the deck." Any breeze, slowed by friction close to the ground, will put up less of a fight if you crawl rather than walk. But when the man had lifted his head and shoulders to see what lay ahead, he gave the wind all the surface that it needed.

"It just flung him away," my colleague told me. "A snow drift at the end of the deck was like a ramp, and I thought he was gone, swept over the edge. We'd have to go pick up his body on the other side of the summit. Somehow he managed to hold on to something, but he wore down a pair of gloves sliding to a halt. It took him forever to crawl back."

Later calculations showed that the man had traveled from stationary to over 30 mph—from Beaufort Zero to Beaufort Six—in a split second.

Commander Francis Beaufort (later promoted to admiral) invented the scale that bears his name in 1805. He wanted to design a way for sailors to easily estimate the power of wind, when no reliable anemometer was available. He based his scale on the flapping sails of a frigate at sea. Calm winds rated a zero. A hurricane shot up to Force 12. Whenever wind blew foamy spray off the white-capped waves at sea, sailors safely estimated winds at above 20 mph, what they called a "fresh breeze," or a 5 on the Beaufort scale.

The scale works just as well on land, using tree branches, flag poles, or chimney smoke. A mere 2 on the Beaufort scale weakly rustles leaves and barely nudges wind vanes, indicative of a windspeed between four to seven mph. Smoke from a chimney drifts east at an angle. Flags flap and flop against their poles like lazy, uninspired patriots.

Even a light flag will not fully extend until Force 3, at eight to 12 mph, when leaves and twigs begin to sway constantly. Force 4 will twirl light, dry leaves in tornado-like funnels across a yard, but the speed of moving air is still only 13 to 18 mph. A stronger wind capable of bending and swaying large tree limbs indicates Force 6: 25 to 31 mph. At 40 mph—Force 8—the branches snap. Telephone and power lines emit sudden, menacing whistles of noise.

Force 10—that is, 55 to 63 mph—will uproot small trees, or snap them at their trunks. The shingles on houses break off and fly. At sea, a wind of equal strength will build waves as high as ships' decks and turn the ocean white with churning foam. A "whole gale" is the seaman's term for Force 10. It is dangerous, but not yet a hurricane.

By the time a tropical storm attains or surpasses the 74 mph (119 kph) threshold, meteorologists rely on a different scale to summarize the damage caused by its winds and by the wet slosh of the sea onto land. The human need to quantify nature in formulas and numbers seems unstoppable. The Saffir-Simpson Hurricane Damage Potential Scale runs the gamut between Category 1 and Category 5, ranging from minimal damage to outright devastation that requires a major evacuation inland. The terrible hurricane of 1938 rated a 3, and that was enough. I intend to demonstrate why.

"ANY STRONG STORM around here has to have at least a little contact with the Atlantic Ocean," says Dave Thurlow. Speaking in the conference room on the summit behind my office, he mentions the example of Hurricane Andrew, which whipped across the Florida panhandle in 1992 and wrenched apart the radar equipment at the National Hurricane Center in Miami. An anemometer on an automated weather buoy in the ocean clocked a peak wind gust at 169 mph—and then its mast bent beneath the fury of the storm and stopped transmitting. The speed of Andrew's highest wind gust will never be known.

The New England hurricane of 1938 took 685 lives and caused an estimated $400 million in damaged property. In terms of fatalities, it cannot compare to the Texas hurricane that wiped out Galveston at the turn of the century—unstoppable even though a frenzied meteorologist named Isaac Cline rode on horseback across the island to warn his disbelieving neighbors. More than 6,000 people died. Most of them drowned in the sea surge.

In New England in 1938, folks knew something bad was coming; no one needed to hop on a horse to spread the word. But few weather watchers guessed exactly how bad. From Long Island to Connecticut to northern New Hampshire rivers flooded, trees toppled, wind roared. Blue Hill Observatory, located just outside Boston, recorded a peak gust of 186 mph. After the storm had passed, much of New England and Long Island lay in ruins.

The hurricane originated in the Atlantic Ocean near the Cape Verde Islands, and it had already reached hurricane force by September 16. Eventually it surged up the East Coast of the United States—much as blizzard-bearing nor'easters do each winter—gathering strength and speed. "The hurricane of 1938 had some extratropical features," says Dave Thurlow. "It had fronts connected to it, which hurricanes usually don't."

The heat of the tropics gives birth to hurricanes; hot air is less dense and therefore rises. "It's just convective energy," explains Thurlow. Hurricanes differ from extratropical systems—midwestern blizzards and Alberta Clippers, for example—which result directly from a clash between cold and warm air. But sometimes a hurricane will "join up" with an extratropical storm and develop fronts. The two storms merge in a powerful marriage of convenience. This union happens only if conditions are exactly right. In September of 1938, they were.

Thurlow calls it "a perfect combination. It was a huge hurricane combining with a very, very impressive frontal system and a ridiculously strong jet stream—it was just perfect."

By the time the hurricane struck New England on September 21, the eye of the storm was quickly moving northeast. "Not only did it have incredibly strong circular winds, but the storm itself was moving fifty to sixty miles per hour forward," says Thurlow. That's fast for a hurricane. As the eye of the hurricane clambered onto land, its own forward speed combined with the circular winds at the leading northeast edge. The wind, already howling, turned savage.

The ground shook, almost as in a myth—a monster from the past had crawled out of the sea and stamped its foot on land. A hurricane had arrived.

MANY YEARS LATER, for no better reason than curiosity, I pull a heavy tome of historical weather data off a dusty shelf to check the impact of the storm here in the mountains. Winds are unusually calm this morning, so I must delve into the past to stir up some excitement.

From the point of view of the observatory summit crew, the 1938 hurricane's position and motion produced a southeasterly wind. The wind's speed rose to the point at which it was strong enough to knock a person down, but it never seriously challenged the world-record wind that had hammered this same location only four years earlier.

Temperatures that day started out at 54 degrees Fahrenheit, a sign of a warm southeast breeze. But the mercury dipped into the 40s after sunset, finally reaching a low of 41 degrees Fahrenheit at midnight after the worst of the storm had passed.

Predawn winds started out slow, in the 20-to-30-mph range—but they kept climbing. The atmospheric pressure, corrected for sea level, dropped to 28.728 inches. As I scan the daily report, I see that an unknown technician with sharp, slanted handwriting has scrawled a message in one corner: "Very high winds during late afternoon and evening with a max of 140 mph for a five minute period and 160 mph for 5 sec at near 500 PM." He also noted: "4.55 inches rain since 710PM last evening. New thermometer broken and replaced at 710 PM." That, apparently, was the extent of local damage.

The hurricane of 1938 was financially and emotionally draining for hundreds of thousands of Americans. Many lost their lives or their livelihoods. But for once, it seems the windy summit of Mount Washington escaped the worst of the blow.

PART III

Extreme Measures

14

Tying Down the Wind

EACH TIME A NOR'EASTER OR A HURRICANE GRUMBLES UP the coast and lashes the mountains with a whip of wind, I start to question my love for the outdoors. It's hard to admire Mother Nature when she's trying to kill you.

When the wind flicks a chunk of ice at your head at 100 mph, you don't much feel like saying "Thank you." In fact, you don't feel much at all.

A young meteorologist once staggered through the door of the observatory and slumped against a wall. His clothes were crusted with rime. Outside, the wind churned the wet air, spinning up a thick brew of ice, sleet, and fog.

The man rubbed his shoulder and winced in pain. "A chunk of flying ice hit me in the back," he said. "It felt like a golf ball, like someone hit me with a nine-iron."

He had walked outside and was instantly swallowed by the blizzard, drowned in it. Like Jonah trapped inside the whale, he sat and shivered in the storm's belly for half an hour, until at last the wind nudged away the clouds and spat him out into the cold mountain air.

Later that day, a local reporter and hiking enthusiast named Ed called me on the summit for a weather update. The strong winds failed to discourage his planned ascent. "Well, I probably won't come up today, but maybe I'll surprise you," he said, wavering.

I was shocked. The idea was crazy. "You'll surprise us, all right," I replied with an exhalation halfway between a chuckle and a snort. "If you show up, we'll be surprised you made it alive."

Winter storms can hide behind blue skies and pounce unexpectedly. Snow and ice drop all at once from the dark veneer of newly formed cloud. The sky emits a howl, like a wolf before the hunt, and then the chase is on. Any stragglers or underdressed hikers caught above timberline must fend for themselves.

Tonight we have been warned; a storm is on the way. The entire crew of the observatory waits in the weather room, safe behind concrete walls. We watch on a small TV as a cold front slithers east into New England, while a snappily dressed meteorologist from Portland stands in front of the map and jabbers excitedly. The cold front is identified by a dark blue line studded with jagged spikes. The pointy side is headed our way, soon to stab the mountains with a sharp bitter wind.

Outside, the sky roars. "The floor is starting to rumble," says Mark Ross-Parent, the bearded staff meteorologist. He walks over to the computer screen and points to a map of the storm. A wing of white clouds flaps over the East Coast, and a fat **L** sits like a brand in red ink over the heart the storm.

Wind always flows in a spiral from high pressure to low pressure. The Coriolis effect—the deflection caused by Earth's rotation—stirs up a whirlpool of wind, a river of moist air swirling counterclockwise around the center of the low. Air surges upward in the center like a spiraling geyser. As the air rises, the atmospheric pressure drops; storms form. Technically, it is imprecise to state that rising air *causes* low pressure. Instead, upper-level winds diverge above the low, creating a suction which makes surface winds converge. The air is indeed rising at this point, but the atmosphere's constant quest for equilibrium continues; air from the surface rises, cools, and actually becomes more dense—heavier! An influx of slightly warmer (and therefore lighter) air helps to offset this effect. Only as long as upper-level divergence outpaces the convergence of winds at the surface will the atmospheric pressure drop. The strongest winds whip around the low nearly 100 miles out from the center, rather than right at the core.

Since a giant **L** is now perched over our neighboring state of Vermont, the sky over Mount Washington begins to howl. No wonder the wind is strong.

The door to the tower emits a sharp hiss as it swings ajar. Upstairs, we hear the outer door clang shut and then a stomping sound echoing on the metal stairs. Finally, crew member Lynne Host walks in, bundled in a blue parka. "The clouds are getting lower," she tells us. "I had to duck." She holds up a hand in mock horror, as if to ward off the sky.

"It's like sitting behind a jet engine out there," she gasps. We all listen; the high-pitched shriek of air never stops or slows. The hiss of wind swells and deepens, a noise like distant tympani hidden below the rocks. That deep thrum of sound is the frantic heartbeat of a storm stirred to rage.

Outside, the sky emits a shrill cry, like a falcon diving to snatch prey in its claws. I glance at the wind chart and see a sudden gust of 135 mph. When I touch the window, I feel the two-inch-thick glass pulsating like a membrane. I yank back my hand, alarmed.

A fierce gust hammers at the windowpanes, and a claw of wind rips off the accumulated rime. Then a gray curtain of fog rubs against the glass and smothers the view; I peek outside into nothing.

Is the mountain breaking apart? Apparently so. The sky inhales a mile of wind in 30 seconds and snorts it through the observatory tower door. Ice pellets swirl in tornado-like funnels and smack against the walls. Off in the distance, a river of blowing snow pours across the summit cone and cascades into a low ravine.

Slowly the mountain crumbles, ducking down through the millennia before the wind's lash.

I DECIDE TO make my first risky foray into the storm at half past noon. The task at hand is to find the precipitation can in the center of the summit cone. Every six hours, a meteorologist must collect the three-foot-long metal can, bring it inside, melt down the snow or ice, and measure the water.

First, to guard against the wind, I wrap my hood tight like a cowl around my skull. A facemask and goggles complete the disguise. Not a single inch of skin stays exposed to the cold bite of the breeze. I walk down the long, quiet corridor to the front door, which has been glued shut by snow and rime. To yank it open I must give a good tug against the ice. The door grunts once in protest and then swings ajar. Blowing snow spills around the edges and drifts inside. The silent air inside the corridor suddenly snarls and stirs. I hesitate.

Just outside the doorway sits a little white hill, sculpted by wind. There is no turning back. I kick an inch of white powder off my ankles and wade through the snow. In the thick of the storm, the clouds have sunk; billows of fog roll in waves across the peak. For a while I step disembodied through this icy fog with no sense of up or down, no direction. A mound of drifted snow trips me, but I see nothing in the way. Is anyone there? White fog pulses in the wind; the mountain has vanished.

By blind luck, I stumble in the right direction. The precipitation can materializes out of nothing, straight ahead. Only a hazy black outline is visible. I anchor my gaze on this solitary, solid object and rush forward before it disappears. Fog gushes into my lungs, the way a breaking wave spills across a ship's deck in a choppy sea.

With one quick swoop I pluck the "precip can" out of its metal nest and turn back to the door. But after I've taken only one step, an iceball the size of a fist shoots out of the fog, punching the can at the edge of my hand. It smacks the metal like a gunshot. Even with padded gloves, I can feel my fingers bruise and numb. The ice shatters on impact.

Is someone out there throwing snowballs at stray Observers? Either that, or spears of ice are breaking off the edge of the tower. A hurricane-force wind hurls them like missiles through the fog.

Soon a second ice block whips past, pitched off the end of the tower. I imagine a baseball pitcher with the strength of Prometheus, standing up on the tower hurling 130 mph fastballs at anything that moves. His aim is off. Ball one.

I crouch down and stagger back to the door, dodging arrows made of ice. My footprints from only a minute ago have already drifted and disappeared, so there is no path to follow. Snow swirls and I am blind. Afraid.

"WE'RE BOUND TO run out of air one of these days," sighs a white-haired visitor, back in the weather room. But so far, we have barely tapped our supply. The lungs of Canada seem never to run out of breath; even two days later, they continue to pump an Arctic air mass our way.

The weather map shows a big **L** perched like a chair beneath a big **H**. "The two pressure systems are right on the edge of New Hampshire, duking it out, and we're caught in the middle," explains Lynne Host. Wind flows like a river from high to low; the steeper the grade, the faster it goes—just as water flows faster over the edge of Niagara Falls than in a flat, sluggish river like the Mississippi.

Exactly how strong is the wind? In many ways, the answer to that question depends on how tall you are. Breezes that curl around a hiker's toes are much slower than the same breeze at eye level. Close to the ground, friction snags the wind and slows it down. Boulders and buildings get in the way.

The strongest wind of the season so far—145 mph—whips past the summit, just as I climb upstairs to clean ice off the instruments. By luck—good or bad, no one is sure which—I get to experience the winter's worst wind full-force. At the top of the tower's parapet, the highest perch for 1,000 miles, nothing shields me from the gale.

I shrink into a ball at the top. With one leg I brace against a post. The wind tugs at the bone in my leg, trying to snap it like a dry twig around the metal base of the post. I clench my teeth. Up above, a ten-foot-tall antenna suddenly snaps off its pole and darts away like a spear.

A chunk of ice slips loose from the far edge of the tower and smacks the back of my head like a flying brick. I cringe until the pain dissolves. Gusts of wind jab my back like the blunt edge of a jack-hammer.

Wounded but still curious enough to test the strength of the wind, I hoist an iceblock over the edge. It bullets into the fog and vanishes faster than a blink of an eye. Is that my fate if I let go?

All solid land is gone, smudged from existence by fog. I could step off the brink and zigzag across the clouds like a kite, floating on this wind. I feel my grip weakening.

The constant gusts snatch the edge of my hood and flip it across my face. The cloth flaps faster than the beat of a hummingbird's wings, a blue blur of fabric an inch above my eyebrows. A cold draft shoots up my back. Perhaps it is time to go. When I grab the rungs of the ladder and swing down, the wind tries to push me over the edge. A breeze claws my back like an icy hand. Suddenly I notice that my windpants have been tugged down to my knees; hurriedly I yank them up again.

Only after staggering downstairs do I notice what other pranks the wind has pulled. For instance, the insides of my pockets are wrenched out, and flap at my sides like elephant ears. An ice-crusted, untied shoelace drags behind me like a snake clipped to my heel. On my head, goggles and hat both tilt, skewed by the breeze. A patch of warm wool from my hat blinds my left eye.

Apparently Mother Nature has a sense of humor. In one fell swoop she groped at my goggles, tore open the zipper of a supposedly windproof jacket, pulled down my pants, and untied my shoes. No wonder the rest of the crew laughs when I walk in, my clothing in disarray, humiliated by the wind.

THE SKY CHANGES moods quickly, and even the most experienced travelers sometimes get caught by surprise. When renowned mountaineer David Breashears visited Mount Washington in 1997, his first response was, "I can't believe it's this bad at only six thousand feet!"

After a trip up Mount Everest lugging a 41-pound IMAX camera, Breashears and his film crew journeyed to Mount Washington to shoot a few minutes of extra footage, avoiding the inconvenience and expense of a return trip to Nepal. "We wanted the worst weather in the world, and we got it," he emphasized later.

"Right after we set up the tents, the fog obligingly rolled in," added one of his film crew, voice muffled behind a black face mask. At one point, the crew member grasped a tent for dear life; the fabric flapped violently in his hands like a magic carpet about to fly away without a rider. A chill wind gusted to 70 and then 80 mph, eventually blowing the parka-clad members of the film crew back indoors, where they warmed their hands around mugs of hot coffee.

Mountains are weather makers. On their snowy slopes, even on the fairest of days the jaws of winter will clamp down on the unwary. The vertical intrusion of a summit into the troposphere forces air to rise and cool; if the dew point is reached, invisible water vapor condenses into clouds. The jump in topography also funnels wind between Earth's surface and the lid of the tropopause at 40,000 feet, essentially speeding up the breeze. So climbers never know for certain what to expect.

In the wake of last night's cold front, the air this morning calms and clears. Sunlight now sparkles on rippled fields of snow and rime. A solitary cumulus cloud creeps east across the Great Gulf, nudged by a gentle wind. The vast blue dome of sky holds no hint of danger.

At 10 A.M., an energetic crew member bundles up securely against the cold and leaves on foot for Tuckerman Ravine, planning to walk back early in the afternoon. For safety's sake, he fills his pack with extra food, clothing, and survival gear that he does not expect to need. Cabin fever has started to torment the observatory staff after a week of solid fog. A long hike in today's crisp, clear air is welcome.

At half past noon, the wind surges without warning. Each angry gust scoops up a fistful of fallen snow and hurls it back into the air. Despite the almost cloudless sky, visibility plummets from 80 miles to 50 feet in a matter of minutes. I have never seen the sky clamp down so fast. We all start to worry about our colleague, but there is nothing we can do. It is too dangerous to go outside.

An hour before sunset, the missing crew member finally walks in from the cold and slumps down on the couch, breathing heavily. His face is ivory white, and he mumbles that his goggles had blown away. His voice is slurred. "I tried to reach over the snowdrift to get them, but the wind hit me right in the face." He shivers and stops talking. "That's the closest I've come to dying," he finally says. "I really thought I was going to die."

A cup of hot chocolate, clutched with bone-white hands, warms up the man. But the story spills out of him slowly, like water droplets dripping off the end of an icicle.

Outside, the windchill has plummeted off the chart, to minus-100 degrees or so. The air is thick with blowing snow. The man explains that he started to shiver as he hiked, and he recognized the onset of hypothermia. Cold and pain racked his body, but he did not dare pull off his mittens to put on another layer. Could he reach the summit in time?

The wind refused to cooperate; it blew harder, pushed him back down the trail. For a while he huddled behind a cairn, half-protected. "I looked one way, couldn't see anything; looked the other way, still couldn't see anything. Then miraculously, the blowing snow cleared just enough to let me see the next cairn. So I sprinted over and waited for the next break in the fog."

Near the rim of the gulf, one of his snowshoes came loose, stripped off his foot by a savage gust. For a second he staggered along with only one snowshoe. Then he kicked it off, let it fly away in the fog.

"My water bottle froze, so I couldn't get that open," he says. As he tells us this story he sinks deeper into the sofa and sips his drink. "When you're two miles from your destination and you can't go back, you can't go forward . . ." His voice trails off. He blinks once and looks up at us, a small circle of openmouthed listeners. But he does not really see us. Instead, he stares deep into his memories.

The man shudders and continues his story. "The wind is tearing at your clothes and you're stuck behind a little cairn, shivering. That's terrifying!"

"THERE'S NO SUCH thing as bad weather, only inappropriate dress," suggests a mountain climber turned meteorologist. This tidbit of weather wisdom comes from bitter experience.

During the winter of 1993–94, a young man named Jeremy crawled up Mount Washington only half-alive. He pounded frantically on the observatory tower's door with his feet; his frostbitten hands had bulged to the size of baseball mitts and were useless. He wore no gloves, just a pair of synthetic palm warmers. Later, when his fingers thawed, they jutted from his hands like rose-red icicles, swollen with blood.

Two miles down the trail near Mount Jefferson, his friend Derrick lay dying inside a cocoon of snow. A rescue team dug his body out of the rime the next morning. "When they freeze up like that," said a rescue worker, "It's like carrying a mannequin. The arms don't move. They're petrified."

"The search-and-rescue teams said it was one of the hardest rescues they've ever done," recalled my colleague Norm Michaels, who was busy treating Jeremy for frostbite and hypothermia back at the observatory. The temperature sank to minus-42 Fahrenheit, with winds gusting above 80 mph.

In such conditions, our frail human bodies act like sieves. Heat leaks through pores in the skin and quickly dissipates in the cold air, whisked away by wind. Wind can snatch heat from our bodies in a fraction of a second, like a thief who has grabbed a wallet and runs off into the night. Sometimes, instead of a wallet, the wind steals a life.

We call it windchill, when the air feels colder than it actually is. When air is calm, you don't notice any windchill. Heat sticks close to the skin like a transparent shirt. A protective layer of warm air extends about a quarter of an inch beyond the epidermis. But as the wind increases, this invisible heat shield quickly shrinks and disappears. Even a soft breeze will make you feel like you have suddenly stepped from a refrigerator into a freezer.

MOTHER NATURE IS obviously a tempestuous lady. I used to picture her as a wiry old woman, her toes dressed in slippers of snow as she tossed snowballs at hikers. She smiled at her victims and her teeth were crooked and yellow, like corn. But her playful grin was only half-friendly; it hid a glint of menace.

The image is almost comic—almost, but not quite. Severe weather—along with reckless behavior and plain bad luck—has killed thousands of people on the peaks of the world. Mount Everest heads the list with more than 140 fatalities. On K2, the world's second-highest mountain, one out of every three climbers has perished. More than 125 people have died on tiny Mount Washington alone.

High winds, killing frosts, and avalanches hunt hikers all winter long, stalking them across the slopes. But not all victims come on foot. Sometimes, a bitter gust of autumn wind will swirl across a summit and rip an airplane from the sky.

In October 1990, a plane crashed in a col near Mount Washington, adding three names to the death toll. No one knew for certain what happened. Perhaps a sudden gust knocked it from the sky. Perhaps the day's freezing fog blinded the pilot or weighted his wings with rime.

At the observatory, the summit crew knew only that a plane was missing somewhere. The craft had abruptly disappeared from radar at 4:44 A.M. Winds that morning averaged 42 mph, with higher gusts. The peak gust for the day topped out at 93 mph. Thick fog buried the summit in rime.

When a hiker discovered metal debris along the trail, the observatory crew put two and two together and feared the worst. Two men grabbed their winter gear and set out from the top to investigate.

"Not quite knowing what to expect, Rich and I headed off into fog, snow, ice pellets, and solid forty- to fifty-mile-per-hour winds," wrote an intern named Doug in the logbook. "Right at Clay Col—signs of the accident—small pieces of aircraft, baggage, etc., on the Great Gulf side of the trail. Saddest of all, though, were the personal effects: a comb, clothes, tapes. It sinks in—these were real, living folks with lives, families, parents, friends."

Bodies lay in pieces not far away. The unedited account in the logbook sickened me; my stomach churned as if I was reading a graphic news flash from Bosnia or Rwanda or some other war-torn hell. And yet the tragedy had occurred just a short walk down the trail from my place of employment. How many times had I stood on the rocks in precisely the spot where three people died? "Quite clearly, we could see where the bodies bounced on the way down. Whatever consolation one can draw, at least they died instantly and did not suffer. Visibility was severely restricted. Perhaps the day's snow and high winds will do a bit to cleanse the Gulf."

Fish and Game authorities removed the bodies the next day, but wreckage from the impact was scattered over the rocks above treeline. Long after the initial salvage operation, small chunks of metal were still being discovered in dribs and drabs.

I am reminded of this fact by a balding park ranger who keeps a jar full of relics from the crash. I glance inside and see twisted metal gears, bits of shrapnel, and one object that looks like but surely cannot be a fragment of human bone. Impossible . . . or is it? I cringe at the thought. "This is a piece of the plane that crashed and killed those guys," he says. "And this is another piece of the plane that crashed and killed those guys."

He pulls out a metal knob and plunks it on the table. Outside in the wind, Mother Nature peeks through the window and grins.

15

The Big Wind

As a web of fractocumulus ripples above the White Mountains, gusts of wind caress the steep peaks with silky ribbons of mist. By late afternoon, strings of fog dangle so close to the summit of Mount Washington that I imagine I could stand on my toes, reach my arm to the sky, and yank down a rope of cloud.

"I think a cloud's trying to sit down on us," says Mark Ross-Parent. He peers out the Lexan polycarbonate window in the weather room, his hands cupped in front of his eyes to block out the glare of the setting sun. Far below the peaks, a sea of stratocumulus boils and churns.

The late-afternoon sun is already a tiny orb far, far away. A brisk December wind howls from the west, stirring the clouds like a giant spoon. Nestled inside the observatory's protective walls, a crew of meteorologists watches the sun slip away. A coil of mist washes against the windowpane and deposits a spray of rime, a sheet of icy pinpricks on the glass. Soon a chisel of wind snaps forward and scrapes the window clean.

To the west, the orange disk of the sun rolls into a mesh of clouds. For a while it bobs just below the surface, a pale globe sinking deeper beneath the waves. As the sun teeters on the rim of the sky, each breath of wind buffets it, nudging it off its perilous perch on the edge of the horizon. One last gust of wind shoves it down, drowns it in mist for good. The sun winks once and disappears. Twilight has begun.

In a far corner of the weather room, the door to the tower rattles and stirs; the invisible hand of wind is trying to yank it open. We can hear the wind whistle and whine through cracks in the tower walls.

I glance over at the Hays chart recorder on the wall, where a strip of red ink indicates a windspeed of 80 mph—a hurricane-force wind. When I touch my hand to the cold glass, the glass trembles, shivering with cold. Fists of wind pound at the windows, groping and plucking off the white rime.

ALL WINTER LONG, the wind cascades like a river across the slopes of Mount Washington. Gusts spear unwary hikers like sharp icicles. Such winds are classified as hurricane-force, but no heat from the tropics pumps their rage. Instead, the orographic lifting of the mountain provokes the wind to hysteria. This is called the Bernoulli effect, whereby air masses tend to speed up as they pass over higher elevations. The jagged neighboring peaks act as a funnel to further intensify the breeze. Swirling air is pinched between high peaks and a lid in the atmosphere called the tropopause, approximately seven miles above the surface—and that "pinch" is enough to speed up the flow.

By coincidence, Mount Washington also sits in a jumble of mica-schist boulders at the intersection of three major storm tracts and air mass routes: up the Atlantic coast of the United States, along the Ohio River Valley, and east along the line of the Great Lakes through the St. Lawrence River Valley. Passing storm clouds drink their fill of moisture from these large bodies of water and later wring themselves dry on the mountains of New England. As a result, the summit experiences hurricane-force winds more than 100 days each year, approximately one day in every three. If the observatory could package miles of wind and sell them for a nickel apiece, it would never run out of supplies.

A solid concrete bunker, the observatory's steel-reinforced walls are designed to withstand 300 mph winds but have never been put to the test. Since the building's construction in 1980, the strongest wind measured has been 182 mph.

Each hour, a lonely weather observer bundles into a parka, scarf, face mask, and goggles and ventures outside to monitor the conditions of the sky. If necessary, he or she also de-ices the instruments at the top of the tower. Such a chore is necessary whenever clouds streaming over the summit deposit rime, a feathery ice that wraps like a cowl around any solid shape, including delicate scientific instruments. These "frost feathers" of rime ice, which can measure more than three feet long, are quite common.

This hour it's my turn to jog upstairs. I glance again at the Hays chart on the way out, just to see how hard the wind will hit me once I poke my face beyond the door. By now, a line of red ink shoots all the way up to the 120-mph mark. The wind exhales, a humming sound like the steady smack of rain during a thunderstorm.

The tower parapet is the highest perch in New England, 21 feet above the summit, with a drop of nearly 65 feet off the north edge. A narrow walkway with a metal ladder extends up to the top, which is deep into the clouds. No wall or shelter shields the upper parapet from the full force of the wind.

As soon as my head rises above the shelter of the tower wall, it bobs like a buoy in a channel of wind. My balaclava mashes against my face in an attempt to smother me, but it needn't bother; in such winds I can scarcely breathe.

Wind slices past my lips like an icy whip. Each gust is so sharp and quick that I barely manage to pull a painful sip of air into my lungs.

I turn around, and feel the hood of my parka meld against the curve of my skull. A savage gust snarls in my ear. How many hundreds of feet of wind slice past the summit each second? If I lose my desperate grip on the top rung of the ladder, how far will I fly before my body falls and breaks among the rocks below?

A pebble of ice hurtles across the night sky and shatters on the edge of my goggles, just in front of my left eye. The impact deposits a starry scar of white powder on the surface of the lens—a pattern much like the craters of sprayed dust left behind when a meteorite collides with the moon.

At last, I pull myself onto the topmost layer of the tower and crouch down, but there is no shelter here, no place to hide. An inch away from my boot, a chunk of ice as large and heavy as a cement block is scooped off the parapet and hurled east, an icy projectile shot straight into the eye of night. The ice block flies away in a horizontal line, immune to the pull of gravity, so strong is the wind.

Invisible fingers of wind slide under my gloves, prying at my fingers, trying to slacken my grip on the handrail. Spears of ice jut from the metal ring, and up above, the wind vane is bundled in rime. It wallows in the wind like a sluggish fish mired in cold mud.

My job is to break all the instruments free. To accomplish this daunting chore, I have lugged to the top the latest in de-icing technology: a sledgehammer with a light plastic head. Four or five good whacks to the parapet's circular ring should do the job. Vibrations of each blow send tremors up the poles, and the instruments shake off ice the way a wet dog shakes off water. At least, that's the theory.

I try to stand up, but the wind pummels my back. It nudges me forward so that my belly is thrust into a spear of rime. Fortunately the ice is soft, and it crumbles; tiny white spheres of rime dash and dart away in the wind, like comets in frenzied orbits. They vanish in the clouds.

I manage to turn my back and hoist up the sledge, but my backswing hits a wall of wind and stops. The air is as thick as concrete. A river of wind flows across the tower with all the strength of a waterfall, an enraged Niagara. How am I supposed to swing—or stand—in such winds?

FIVE MINUTES LATER, down in the concrete bunker of the observatory, I peel off my hat, feel the electricity in my hair mingle with the air; I probably look a bit like the famous picture of a frizzle-headed Einstein. But I don't feel half as smart. The wind nearly won tonight's battle.

Exhausted, I pull off my gloves, the outer shell of my windbreaker, an inner coat, double boots, and windpants. My entire wardrobe accumulates in a mountain of fabric on the table. An arsenal of winter garments is essential whenever the breeze stuffs the windchill factor off the bottom of the chart, making the air cold enough to freeze exposed skin in a fraction of a second.

Outside, the wind is a sustained howl, but Mark Ross-Parent dismisses the noise with a shrug. "This is nothing," he says. "You can really tell when it's blowing a hundred and fifty miles per hour out there. You don't need to look at any instruments. You don't need an anemometer. The wind is a presence, rapping at the windows. Trying to get in. Trying to get you."

Sixty-five years ago, a technician named Sal Pagliuca didn't wait for the wind to come find him; he walked outside and willingly flung himself into the strongest wind ever recorded on Earth's surface—231 mph.

Pagliuca was a dark-haired man from Italy, one of four original employees of the Mount Washington Observatory, which was established in 1932 to facilitate weather observations, scientific research, and educational programs on the summit.

When an astonishingly swift river of wind surged across the mountain and slammed against the walls of the small wooden shack where Pagliuca was working, neither he nor his co-workers knew what to expect. They did notice that the cat, Oompha, preferred to sit close to the fire instead of at her usual perch by the window—and no wonder; the window bulged inward an inch and a half, bending to the brute strength of a southeast wind.

It all started on a seasonably warm day for April, with temperatures in the 20s and several inches of wet snow falling from the sky—nothing out of the ordinary. But wind gusts soon topped 100, 150, 175 mph, still rising as the morning advanced toward noon. Rime ice covered windows, walls—and instruments. As chief observer, Sal Pagliuca later wrote in the observatory logbook: "Mac, Steve, and I had been taking turns in knocking ice from the anemometer post. Art and George our guests made themselves useful in many ways but particularly in attending to the fire."

During the peak of the storm on April 12, 1934, Pagliuca risked a trip outside to de-ice the instruments. It was the old roof of the first observatory building, a narrow wooden box buried under rime. The walls shook, trembled, and flexed in the wind—but they held. No one blew away.

In the logbook, Pagliuca describes his experience, climbing up the ladder to the top. Halfway up, he wrote, he made the mistake of glancing back at his colleagues, who were filming his climb up the ladder. In doing so, he turned to stare directly into the wind. Like a deer blinded by bright lights, he froze.

"I felt the full force of the 200 mph SE wind on my face," he wrote in the logbook. "It was the warning that I had better proceed for the anemometer."

Half a minute later, he had reached the top and hammered at the ice. "Kneeling on the platform I pounded the foot-thick ice accumulation with all my strength. But alas! I was not making much impression on the rough frost. Perhaps a sledgehammer would have done a better job but I doubt if the strength of Polyphemus could move a sledgehammer in a 200 mph breeze."

The spidery black ink of his log entry is difficult to read in places. He continues: ". . . and I thought of mighty Prometheus as the wind was furiously blowing my parka out of my storm pants, the hood on my face almost blinding me. I could not waste much more time on the roof . . . I reclined for a few seconds against the ice column, then I must have yelled something muffled by the leather mask, the hood, and the roar of the wind, as I jumped up and pounded the ice again. *Urra* Big chunks of ice were now flying northwestward and disappearing in the dense fog without ever having a chance to bounce on the roof."

Down in the observatory, a pair of men with stopwatches were timing the clicks, eventually measuring a 231-mph gust. "Steve was timing gusts of 220 mph and occasionally of 229 mph. As I returned inside I could not believe what Steve told me, so I took the stopwatch from him and started timing gusts myself—Yes 229 mph—once— twice—then a lull— . . .

"Our first thought was, will anyone believe us?" Pagliuca wondered.

The men anxiously watched their walls and windows bulge under the impact of gusts. Their freestanding building, latched to the boulders by heavy chains, yanked and tugged at its restraints, like Prometheus in torment on his rock and struggling to break free.

The youngest member of the crew, a 26-year-old engineer named Wendell Stephenson ("Steve") had also climbed out to de-ice in the early hours of the storm. But he was pinned to the ladder by a 160 mph wind, unable to climb up . . . or fall down. "If I had known how strong the wind was," he said later, "I'd never have gone out there."

In a wry understatement, radio man Alex McKenzie later wrote, "The gusts of wind against one's back were not caresses."

Something special had happened in that storm, without a doubt. But were the measurements correct? Was 231 mph an official record? Months later, the United States Weather Bureau approved the results. The anemometer proved accurate, and a new wind-speed record was penciled into the books.

A NUMBER OF questions have been aired in the years that followed the measurement of the world-record wind. What was so unusual about this storm? How did it produce such an unexpectedly powerful blast? Why has the record never since been equaled or even approached on Mount Washington? Will another gust ever top 200 mph?

In search of answers, I decided to check with Dave Thurlow, host of the syndicated radio program *The Weather Notebook*. On his desk in the office of the popular radio show sits a map of the eastern United States with a weather map of April 12, 1934, superimposed on it. "One rainy day I guess I got bored and plotted the data from the Big Wind," Thurlow explains with a shrug.

On the map, the spiked blue line of a cold front trails off to the southwest, while Mount Washington—"the Rockpile"—sits near a northward-moving warm front. Curiously, wind speeds at stations closer to sea level are downright slow, at least by comparison. "There weren't a lot of really big winds recorded at low elevations like Boston,

Portland," says Thurlow. Portsmouth, New Hampshire, reported a windspeed of 50 mph. Portland's wind came from the east at 45 mph, a mere 8 on the Beaufort scale. Breezy, perhaps, but hardly strong enough to knock down a tree, much less a world record.

Conditions obviously favored stronger gusts on Mount Washington. "We were definitely in the windiest part of the storm," says Thurlow. But why?

Normally, weather systems move west to east. They flow in the direction of Earth's rotation and—more or less—follow the path of the jet stream. A typical storm will push east over New England and then move out to sea, never to be heard from again. But the storm of April 12 was not typical. Before it swung out into the Atlantic, the storm halted its eastward progress and headed back to the west, like a would-be swimmer who suddenly decides the water is too cold and walks back up the beach instead.

"It's not unheard of," Thurlow comments. "It's the type of weather pattern that can produce big storms." A weather textbook will describe the process as "retrograding." That's what occurred during the deadly October 1991 storm chronicled in Sebastian Junger's book *The Perfect Storm*.

In the case of the storm that gave us the Big Wind, low pressure originated in the Ohio Valley and tracked to the east, to the coast of Virginia. There, it turned to move up the Hudson River Valley into New York—which is where it was centered on April 12, while Sal Pagliuca was busy de-icing the anemometer on Mount Washington. Later, the storm turned sharply again and swung into Ontario.

The cause of retrograding was a twist in the jet stream miles overhead. The pattern of the jet stream was a curve, a shape much like a horseshoe that opened to the north. The storm simply followed the arc. "It came up the coast and basically took a left," Thurlow explains.

The cold front had not yet reached Mount Washington, but a warm front—a clashing of cold and warm air masses—was already advancing. "A warm front comes through the area—with warm air overriding cold air—so that higher up, at two or three thousand feet, you have high winds," says Thurlow. At the same time, gentler breezes sifted through the valleys. It's not an uncommon scenario, he explains. "I've seen days where the summit could be one hundred or one hundred and fourteen miles per hour, and the valley would be calm."

The warm front continued to move north—a process called "overrunning"—and the southeasterly upper-level winds pouring in ahead of the front were what eventually produced the 231-mph world-record gust.

WHEN THE GUSTS finally dip back below 110 mph, I hurry upstairs to make another attempt at de-icing the tower. A surge of air pressure keeps the top door shut; I yank it open, unglue it from the sticky air.

A westerly wind still growls, but it has begun to veer into the northwest. Immediately it snatches me by the shoulders, swings me around, shoves me away. My stomach thumps against the metal ring around the parapet and I spit out a gasp—a tiny, futile sound, lost in the thick of the hurricane like a tiny pebble vanishing with a delicate, unnoticed plop into the roar and surge of a waterspout. This time, however, I am prepared for the blast of wind. I grit my teeth and succeed in de-icing the instruments without too much trouble.

Wind travels in swirls and eddies, and its strength increases with exponential force. "As wind speed increases, the force of the wind increases dramatically," explains one of my colleagues to a visitor later that night. "A wind that doubles in speed, say from ten to twenty, isn't twice as strong, it's four times as strong."

I know this already from personal experience, and have the bruises to prove it. More than once the wind has shoved me flat to the rocks with a thud.

One day months earlier in the thick of winter, I pulled open the door at the top of the tower and stepped outside. On the ground I saw a man, bundled in fur, crawling in slow motion across the deck.

It was our cook, Ira, buffeted down to his hands and knees. The wind had snatched him by the neck, twisted him around, and made him pay homage. How long had he been out there, a stone's throw from the door but unable to reach it?

The thermometer next to the door was crusted with rime, but I scraped off the tube with a fingertip and traced the silver line down to the minus-14-degree Fahrenheit mark. All around me the wind roared like an engine, unstoppable.

It isn't possible that Ira heard me arrive on the deck, but still he lifted his head when I stepped outside. That extra inch of forehead, when he glanced up, was enough of a surface for the wind to knock him backward. He bent down again and struggled to regain lost ground.

Except for the brown parka that ballooned around his waist in the wind, he looked just like a shipwrecked sailor washed up on the beach, too weak to crawl free from the taunting slaps of the waves.

After a struggle, he finally reached the lee of the tower; the wind released its icy clamp on his back and allowed him to stand. He staggered over, slipping on the snow. His goggles were frosted. He leaned and shouted in my ear.

"What?" I shouted back, unable to hear above the roar. He tried again, but whatever sounds issued from his lips slipped away in the wind before they reached my ears.

"What?" I roared back a second time.

He screamed, and I finally heard a muffled whisper siphoned through the hurricane breeze. The pale shadows of words, as if echoing up from a deep well, wiggled through the gusts. "I said," he gasped, "you can really tell the difference between eighty and a hundred miles per hour!"

I nodded. What more could I say?

Downstairs a few minutes later, I heard a colleague describe the wind as "a beast trying to break in." Yes, we all agreed that it was a beast. So why did we go out and deliver ourselves to its jaws?

I thought about the words of Wendell Stephenson, who lamented back in 1934, "If only I'd known how strong the wind was . . ." But he did know, of course; he had heard the shrieks, had seen the rime-crusted glass windows ready to implode.

We all had known, had seen the magic number "114 mph" on that day's wind-speed chart. But did we really know what that number meant? The wind cannot be quantified; a number is not enough. Nor is it enough to see the walls bulge, straining against the invisible weight of the wind. Perhaps we can never truly know the strength of a hurricane until we feel fingers of wind clamp around our necks like icy talons and hurl us in humility to our knees.

16

Dancing the Tango
with Mother Nature

TONIGHT, THE WIND STAYS SILENT. THE SKY INHALES A SIN-gle breath and holds it all evening, waiting for dawn.

On the ground, the snow makes a powdery sponge that absorbs sound. Green needles of spruce trees bristle and shiver inside a lattice of white crystal. The sky is as quiet as a sheet of glass.

WHEN SUNRISE STARTS to warm the December air, the silence is punctured at first by the beak of a pileated woodpecker, tapping in the distance on the knobby bark of a pine.

No one else is awake to hear the woodpecker. The rest of the world is asleep, unaware; even the sky's eyes are shut, crusted with snow. Off to the west a silvery full moon pales in the sky. No one has told it yet that the sun has finally risen; it still slumbers on the horizon, forgetful that the laws of physics require it to make room.

At dawn, weather observers Mark Ross-Parent, Lynne Host, and I bundle in three layers of wool apiece and set out from the base for the summit of Mount Washington, a brisk hike to the observatory where we work. I bend to retie the strap on my crampons, which have wiggled loose. The others pause along the trail, looking for the wood-pecker. Lynne steps off to one side and peers through the trees. Her boots dig a trench in the snow behind her.

We find the woodpecker perched halfway up the trunk of a middle-aged pine, its talons clenched in the bark. The bird's eyes glisten like moist berries in the sunlight. It pays us no heed. Patiently, it taps a message in Morse code on the bark, a steady rat-a-tat. Its beak chisels holes in the quiet air. Invisible sound waves ripple toward our ears.

As Lynne steps closer, her boots crunch in the ice, a noise loud enough to warn the woodpecker. It stops its work and listens. For a minute that feels like an hour, we are all silent and still. The wind is asleep. Nothing moves to mark the passage of time. Earth's gears grind to a halt, each sprocket jammed with ice.

A single pinecone falls from the treetops, knocking a tiny crater in the snow. The cone smacks the hard-packed ice with a sound like a distant cough. And that lonely noise is enough to set the planet spinning again.

The woodpecker returns to life, but only for a moment. The swift drumming of his beak pokes another hole in the air, a breezy cavity that floods with silence. When we do not move away, the bird stops again to watch us. It is so quiet now that I can hear the trees breathe, their bark cracking and snapping with cold. Just up the trail an old hemlock bends and sways, like an early riser stretching his limbs.

The air crystallizes, a transparent sheet of ice. I am reluctant to move, to impose noise on the still air with the crunch of my boots in the snow. All around, the spruces and pines bristle on the mountain's slopes, their sagging branches freshly ornamented with bracelets of ice.

Up above, the sky purples. A breeze starts to dab my forehead like a damp cloth. Together the three of us continue to plod up the mountain.

In the distance, fog drapes across the shoulders of Mount Adams. And just below the peak of nearby Mount Jefferson, a clump of cumulus fractus hangs in the still air like a tuft of beard, its fibrous hairs beaded with ice.

A raven hovers above us on a pale breath of wind, while Earth rotates below. He has only to wait a moment to pass from mountain to mountain, state to state. If he is patient, the Green Hills of Vermont will roll under soon enough.

Such a day is when winter smiles. Miles up in the sky, a few thin cirrus clouds drift like feathers toward the sea, a stream of ice crystals kept far aloft, where the air is too thin to be breathed. The clouds refract the sun's rays in a quasi rainbow. A pair of sundogs—optical illusions, twin suns—grin like dimples next to the sun's face. Faintly colored, these sundogs are also known as parhelia, or mock suns.

Cirrus clouds are ice crystals high in the sky, usually well above 20,000 feet. Sundogs are the refraction of light through the six-sided crystals, the prisms of snowflakes that stream across the sky as cirrus clouds. Ice crystals in clouds producing these optical phenomena must be oriented vertically, just so. Otherwise, if crystals tumble at random in the high-altitude winds, the full circle of a halo might surround and enclose the sun. At times, a complete ring of white or reddish light orbits the daystar in an apparently clear sky, a sign that cirrostratus clouds are far aloft. Cirrus is snow that never falls to Earth. Angel snow.

Upper-level clouds often scout the sky ahead of low-pressure centers. They are letters of intent from a coming storm—omens. "The bigger the ring, the nearer the rain," is one expression. Also, "When the sun is in his house, it will rain soon." The halo is the key. But if this morning's temperature is any indication, we can expect snow instead of rain. Another clue to Mother Nature's mood is sometimes provided by wind that shifts direction with altitude. Simply by watching the motion of clouds, it's possible to determine which way the wind is blowing at higher altitudes. "Veering" refers to a clockwise shift in the wind direction; "backing" is a counterclockwise change. If ground-level winds blow from the southwest, but a middle-layer of altocumulus appears to drift westward across the sky while high cirrus push northwest, that is a sign of winds veering with altitude—and usually heralds the arrival of a warmer air mass. Backing winds with altitude mean that colder air is advancing. Based on what I can see in the sky, I now expect a warm front to push through the region later today.

Foul weather is on the way. We have been warned. Winter has not forgotten the mountains; it is soon to return.

———

IN THE MOUNTAINS, SNOW FALLS IN LUMPS of hundreds of inches per year, and blizzards stirred by the wind suffocate stray hikers with swarms of biting, blinding ice crystals.

By early afternoon, our hiking expedition encounters the angry face of winter. The sky frowns, and snow spills from its downturned lips. At first, just a flurry of snowflakes parachutes down. It soon settles into a steady downpour of snow.

Wind prods the ground in white coils, playfully scooping up a backyard's worth of snow and hurling it aloft. It turns mean-spirited and flings snow at our faces; the ice glues my eyebrows in place, burns my skin like cold fire. I turn around to escape the barrage, only to discover that the trail has disappeared behind me, erased by wind and snow. We cannot turn back.

Year after year, the blunt edge of winter strikes the summit like a mallet. Winter is a hammer swung by a strong, invisible hand. It splits boulders into rubble, chops mountainsides into neatly packaged loads of pebbles and clay that are light enough for rivers and rainwater to carry to the sea. Tentacles of wind poke the ground, stuffing ice into nooks and cracks. As millennia pass, boulders crumble; mountains wash away, back into the belly of the Atlantic.

The promise of morning has been swept away by the breeze. I reach up to pry the ice off my eyebrows; it melts away, squeezed like cold juice between my fingers. The sleeves of Lynne's blue parka turn ghostly white with rime.

White fog pulses in the wind. Each gust fingers and fusses with the zippers on our jackets. The wind gushes into our mouths, and one sudden gust penetrates my hat and playfully tweaks my ear. Already, Mark's beard hangs off his chin like a frayed cobweb, heavy with ice.

We don't belong here in the cold; this is no place for the living. The fog and wind erase us, paint us white with thick slabs of rime. Rime blots out our blue coats and orange caps, makes us meld with the winter air. As we hike, our footsteps vanish behind us. Is an eraser swiping at our heels? Only the stubborn muscles in our legs propel us forward, keeping us just ahead of the groping hands of rime. But how long can we keep the pace?

Off to the side, barnacles of ice cling to the rocks. The summit looks like a stark island newly risen from the waves of clouds. And so it is; the cycle of winter is renewed, and by season's end the dust and dirt of summer are always scoured away.

A few yards up the trail, Mark's boots kick up a cloud of snow. The summit is nowhere to be seen, so we stumble blindly, trusting to luck that solid land will always be there to accept our next forward step.

We are not taken unawares. Winter is no stranger here. I think back to a warm night in August, many months ago, a night when the cold breezes of winter first swirled across the summit. That night, I learned for the first time that weather is never just an icy breeze, or a cold rain, or a hammer of wind. It is more than just a hindrance or an obstacle to walk through, a drift of snow to shovel.

"WE'RE IN A temperate zone," explains Mark Ross-Parent, "but we're really an Arctic island." It is August again, and though we don't know it yet, the first night of winter is soon to begin. I glance out the window of the observatory in anticipation of the season's first snowflake.

Cold drizzle dresses the boulders in a clear crystal sheath. At 8 P.M. the red ink on the thermograph suddenly swings south. The atmospheric pressure plummets, as if some giant in heaven was drinking up the sky. A low pressure system called an Alberta Clipper lurches into the northeast quadrant of our weather map. The direction of the wind swings from south to north. Hour after hour the temperature drops, until it finally settles into a horizontal line at 32 degrees, the freezing point of water. But still no snow rewards our patience.

"It looks like we've run right up to the brink of winter but turned back," says Lynne. "I guess we're not ready yet."

She is wrong. Less than an hour later, I trudge up the metal walkway to the tower exit to take a peek outside. Wind roars, but I can see nothing from the edge of the doorway. Light gushes out of the tower through the door, spilling into the night. I walk outside, straight into the grip of the wind.

Only now do I notice that I am still wearing tennis shoes. And why not? Earlier today it was sunny and warm; the rocks were dry as old bones.

When I step across the deck, the 80-mph winds nudge me away from the door. My shoes skate across a half-inch thick layer of glaze ice, a slippery cement put in place by hours of freezing drizzle. I can't stop; even when I kneel to get out of the breeze, the wind pushes me on. Like a hockey puck, I slide away.

In the distance, the railing approaches in slow motion. Beyond it lies a twenty-foot drop, and death or injury. I grope for the railing, but my hand closes on wet ice and slips free. I reach one last time for the bar, feel my hand slide along the ice until at last I slow to a stop.

Wind swirls past my shoulders, as strong as a firm hand trying to push my head to the ground. Is there no way back?

I turn, try to stand and run for the door. Though I lean into the wind and sprint forward, my legs chop uselessly on a treadmill of ice. Slowly I drift backward again, back to the brink.

I have learned something tonight: tractionless tennis shoes don't work well on ice. I make a mental note to mention that fact in the logbook when I get back inside. *If* I get back inside.

Finally, I lay belly-flat on the ice, presenting as small a target as possible to the wind. My gloves, thankfully, provide better traction than tennis shoes; hand over hand, I pull myself across the slippery cement, closer to the safety of the tower door.

A continuous hurricane-force wind hammers at the summit. But at last I'm safe again. I stand in the lee of the tower, just inches away from gusts that could whisk me away like a fallen leaf in a stream.

DOWN BELOW TIMBERLINE, many miles away, such a ferocious wind would snap the trunks of trees like pencils and rip roofs off houses. "Such winds rarely occur on land," a page in a popular weather book informs me. But on Mount Washington, hurricane-force winds occur in every month of the year.

Weather watchers are spoiled by all this wind. Our perspective is askew, ruined by riches. The observatory crew makes sport of the powerful blast, playing games with destruction. We grin with feverish enthusiasm. Our eyes shine with an eager glint, like the happy eyes of TV meteorologists who fall in love with blizzards and hurricanes.

Late tonight in the safe bunker of the observatory, we all sit in our chairs watching the Hays chart as the wind surges. A sudden spike hits 97 mph, then nudges past 99. We wait. Will it crack 100?

Gradually, the winds weaken, dropping back to 65 mph, a strong gale. As the clock ticks down to midnight and pushes the minute hand into tomorrow, I jot a quick note in the logbook: "Temperatures continue to plummet, but the big wind that the weather service promised us never shows up."

In the margins, someone else has scrawled—it looks like Mark's penmanship—"Well, we did hit 99 mph. How jaded we've become!"

The next day, the handwriting changes once again as someone else quips, "Mere hurricane-force winds are so dull."

In snow, in winter, the wind takes new life. Even in a mere 45-mph breeze, the air becomes a white sheet, an icy blindfold. One clear day I watched drifting snow cascade up the slopes like a ferocious whitewater, a Colorado rapid running uphill, lashing at the boulders.

During the regular observatory shift change after a long eight days on the summit, the crew rides down in the Bombardier snow tractor, a tank-like treaded vehicle that crawls down the slippery slopes. Force of habit makes us call the machine a "SnoCat" or sometimes just "Cat," though that is a brand name and, in this case, incorrect.

Today, visibility is near zero. Chris Uggerholt, the Bombardier operator, grumbles that he can see just as well with his eyes shut. Wipers scrape against the windshield, swiping at the flurries of flakes like slender arms trying to ward off a swarm of bees. As the Cat lurches forward, its treads etch a waffle-grill pattern in the ice.

In back, five passengers sit huddled against cold metal walls. The windows are frosted; we see nothing outside at all, no hint of the

2,000-foot drop into the Great Gulf that dangles off one side. No one guesses how close we are to the plunge.

"Stop!" shouts one of the crew, riding shotgun up front. "We're going off the road." For a few minutes he and Chris confer in hushed tones, then turn their heads simultaneously to the back.

"We can't see the road," they explain. "Someone needs to go walk in front."

Well, why not? I wonder. I slip back into my balaclava and face mask, plunk a set of amber goggles over my eyes. Then I get out to walk in front, to follow the road. It seems I'll walk down after all, while my comfortable ride tags behind. Mark joins me, and we step along opposite sides of the road as well as we can.

For half an hour we jog in front of the tractor while waves of snow lap waist-deep at our legs. The wind coils and blinds like a gauze. I step forward, and my boot disappears. Suddenly my shin hits a boulder.

What is a boulder doing in the middle of the road? Or rather, where did I go astray? Where did the road go? Prickles of rime multiply on my goggles like a pox; I swipe them away with a glove and continue searching for the correct path.

Mark holds up a hand to warn the tractor away from the precipitous drop, then traces our footprints—what little remains of them—backward. The road sits a few yards away, buried under a drift.

For a time we play this game of hide-and-seek in the blowing snow. I imagine that the road is moving, a slithering snake trying to elude us. But we stomp down on its tail and hold it still; the Bombardier finally crunches forward to safety.

Five miles from the bottom, a curtain of snow opens to clear air, and the flakes now fall in straight lines, free at last from the dangerous tugs and pulls of the mountain breeze. We have arrived.

17

Adventures in Weather

SPRING AND SUMMER OFFER A DIFFERENT SORT OF WEATHER. The change of seasons makes hikers wary of new hazards on the upper slopes: rain-slicked rock faces, sudden downpours, and thunderstorms provide the usual dangers. Melting snow bridges and falling ice can still surprise the unprepared during the transition between the two seasons. As a public-service bulletin at an alpine ski trail once noted: "Holes in the snow have been known to appear and swallow the unwary." And sometimes we meet obstacles that no one expects at all . . .

"THIS IS AS nice as it ever gets," I announced on a late March morning at the base of the Mount Washington Auto Road. "It shouldn't take very long to get to the summit."

Four hours later, I was still eating my words. Wind-sculpted snowdrifts buried the road and nearly defeated the combined efforts of the TV-8 and Observatory plows. Getting down the next day proved even more of a challenge.

After reading Jon Krakauer's exciting tale of mountain adventure, *Into Thin Air*, I've decided to call this episode *Into Deep Slush*. Here's what happened.

The distant summit was white with snow and rime on the morning of March 26, 1998, but a warm southwesterly breeze in the valley heralded the start of spring. I squinted in the sunlight and glanced up at a blue sky decorated with wispy cirrus clouds. No clouds obscured the summit—yet. A nearby river surged and flooded with meltwater. It was so warm that I expected dandelions to pop up through the snow at any moment.

My purpose was to teach a class, or "EduTrip," on the summit, appropriately titled "Life at the Top." Up one morning, down the next afternoon. Given the calm weather, I didn't expect any problems with transportation. (A week earlier, I had taught a class on the topic of "Wind," and, predictably enough, we experienced no breeze stronger than eight mph. Murphy's Law. I told the group to sign up next time for the trip called "Calm.") At the start of this week's trip a group of seven participants and two trip leaders waited at the base of the mountain, bundled in parkas, coats, and plastic boots, and carrying ice axes and ski poles—all the esoteric equipment that the winter clothing list requires. Despite the summery weather, our group looked ready for a dogsled race to the North Pole. "We probably won't need all this stuff today, but you never know," I apologized.

"I hope we get to see some foul weather, too," said one of the group.

The Bombardier snow tractor traveled quickly over the first four miles of road, until we stopped for a camera shoot at the base of the winter cutoff, just above tree line. Even then, the weather was fair. The wind barely whispered. It was warm enough to peel off a layer or two of wool. But the cirrus clouds were mostly hidden. Fast-moving clumps of altocumulus whisked across the sky. Obviously, the wind had started to howl at higher elevations. A gray fog bank hovered ominously just above the highest peaks.

The snow turned dense and heavy, and plowing became a chore. As trip co-leaders, Jennifer Morin and I walked ahead of the tractor at one point, taking the EduTrip on foot over the drifts. Jennifer, a thin, athletic woman with short-cropped hair, led the way; I "swept" from behind, making sure the group stayed together. At a sheltered twist in the road called Cragway, we waited until the two vehicles had plowed the five-mile grade. (WMTW TV-8 in Portland, Maine, maintains a transmitter building on the summit that is occupied by two technicians year round, and it has its own source of transport.) The wind began to hiss and bite. We took refuge from the wind behind a wall of snow-covered rocks.

It was four and a half hours before we finally reached the summit that day, but the real excitement occurred on the way down the next morning.

Chris, the tractor driver, called for an early departure. The temperature had risen to a balmy 41 degrees, so he expected a river of meltwater pouring down the Auto Road with a slick layer of ice underneath. "If we get into trouble, I want to make sure we have time to get out of it," he informed us.

We departed the summit a few minutes after 11 A.M., never suspecting we would not reach the bottom until six hours later.

Less than a mile from the summit, passengers felt the snow tractor suddenly lurch to the right. The right side sank into melting snow, like a ship taking on water. A side window slanted down and gave us a clear view of a cold river gurgling and rising around the edge of the tractor. The window was open a crack for air, but we quickly shut it before the unexpected "river" could flood the inside of the cab.

Tractor treads spun noisily on the ice but found nothing to grip. We weren't going anywhere for a while.

After some heroic shoveling efforts, Chris had diverted the Auto Road River down into the Great Gulf, where a glacial tarn called Spaulding Lake undoubtedly got a little bigger. One of the summit crew started to hike down with extra shovels and a thermos or two of hot drinks.

In the meantime, all we could do was wait. Who could have guessed that one day it would be possible to kayak down the mountain—on the Auto Road, no less? Someone even quipped that life rafts attached to the Bombardier might be a good idea in the future.

A man named Paul from TV-8 came to the rescue an hour or so later. He navigated the WMTW Piston Bully up a slushy, slippery road, but was still a mile below us when the Auto Road River stopped his progress. Suddenly we had a choice: walk up, or walk down? I supposed, if worst came to worst, we could always swim.

Winds gusted above 70 mph and made a hike in either direction somewhat challenging. The group opted to go down. To make the trek easier, all travel bags, sleeping bags, and other cumbersome paraphernalia were stowed in the Cat.

In single file, we walked down the road to where the TV-8 snow tractor idled. Wet feet were the inevitable result; everyone sank knee-deep in slush on occasion. One participant even lost a leg—temporarily. We dug her out and continued on our way. The snow was soft and deep. And all the while, a brown river surged and flowed around us. Fortunately, a firm ridge of snow on the high side of the road made the hike bearable. Since the topic of the trip was "Life at the Top," our unexpected adventure proved educational as well.

"We got more than our money's worth," commented one of the participants at the end of the long day. All in all, the trip offered a little bit of everything in the weather department, from sunshine to snowdrifts to slush.

Snow continued to melt at an unprecedented rate. More than 30 inches had disappeared from the snowstake overnight. We noticed on the way down that Cragway—completely buried in snow only a day ago—was now muddy and bare. So much for winter.

WHERE DID ALL that water go? It was a question I asked myself shortly after returning to the valley. I was already familiar with the phrase "a following sea." But I never expected to encounter "a following snow" in the town of Gorham.

Record-breaking temperatures in New Hampshire—and across New England—burned away mountain snowpacks practically overnight. All that water had to go somewhere, and much of it flowed locally into the Androscoggin River, which flooded its banks in Gorham, where I lived. Sections of road vanished underwater. The parking lot at Mr. Pizza (the favorite restaurant of off-duty weather observers) turned into a pond.

Floodwater in the valley turned me back a few days after our ill-fated adventure. I couldn't drive across town. My boots were still wet from the trip to the summit, so I didn't relish the idea of turning my car into a boat.

The great irony is that some of that water had originated as snow in the Presidential Range. Perhaps it all started to melt on the very day that seven soggy visitors were hiking through slush on the Mount Washington Auto Road. I wouldn't be at all surprised.

All that water gushed down from the rocky upper slopes, carrying the salts and minerals of the mountains down a string of rivers to concentrate in the vast ocean—which explains why the oceans of the world have steadily gotten saltier over the eons of geologic time. A portion of the meltwater I waded through on Mount Washington that day would later flow into the Atlantic. The sun on the ocean would then evaporate that same water; wind would push it ashore somewhere, where it would fall again as rain or snow. And so the water cycle that perpetuates our weather continues.

TOO OFTEN YOU hear a storm before you see it coming; the wind knocks you flat, drives you indoors. But even the sheltered interior of a building is not impervious. The breeze penetrates windows, doors, and cracks to find you even between the walls.

Once, in July, an unexpected gust nudged open a window in the observatory and swept up all the papers on my desk. METAR codes and to-do lists flew. The room filled with scattered rectangular leaves.

The forecast called for a typical summer rainstorm, but so far only a sputtering drizzle wet the windows. As I gathered up the papers and forms, however, a bellow of thunder issued from the clouds; the sky announced its anger in no uncertain terms. I never saw any flash or flicker of lightning. The noise came out of nowhere—out of the floors and walls, it seemed. For a few breathless seconds I waited for another jolt. I half-expected to see an angry Zeus pounding on the windows, tugging at his beard, demanding to be let inside.

A second passed, then two, then 20. Sheets of rain gushed across the naked rocks and drowned the boulders in a slick, wet sheen. If Zeus was really out there, I hope he had an umbrella.

Why did that blast of thunder surprise me? Earlier that day, I had seen towering cumulus clouds bulge vertically in the sky and at times drift across the summit as fog. But until that moment the showers and drizzle were light and intermittent. The fog stymied my vision, concealing the dangerous six-mile-tall crest of an anvil-shaped monstrosity looming above the summit.

Thunder boomed across the mountain again; I jumped. Tremors from the explosive peal shook the floor of the weather room.

WE ARE MADE of water; the human body contains approximately 40 liters of the precious liquid. Water is one of the three crucial ingredients in weather (sun and dirt are the others) and appears in many forms: an anvil-headed cumulonimbus cloud poking into the stratosphere, a river of slush on a mountainside, a softball-sized hailstone, a delicate snowflake, beautiful frost feathers on your windowpane in autumn.

To the inhabitants of the land among the clouds that is called Mount Washington, the most familiar form of water is fog. Fog seals us in a dimly colored world all our own. It surrounds the summit an average of 300 days a year. Moist air surges up the slopes, cools to the dew point, and condenses into cloud. As long as the summit refrigerates in the cold embrace of orographic clouds, we see the world only in shades of gray.

Tonight, as usual, the peak is obscured. As night thickens, the humid air condenses until visibility plummets to zero. "I'm going out to inhale some fog," announces Lynne Host, walking into the weather room. Her voice is muffled by a blue balaclava pulled over her face. A pair of amber ski goggles completes the disguise. She now looks more yeti than human, protected from the icy bite of wind. Her words emerge faintly through layers of warm fabric, like a voice echoing up from a deep well. Gusts howl; the door rattles in her hand.

Even though it's September, late summer in the rest of the United States, a crust of rime already covers the windows—a blank white wall made of frozen cloud. There is only a palm-sized gap in the glass, where the wind has pried open a peephole. I peer outside and see the misty sky drape black curtains over the mountain. Dark fog gushes in waves. Up above, the last stars dim.

Lynne clutches a flashlight in one gloved hand and wrenches open the door with the other. There is a sudden exhalation of wind through the rooms of the observatory. I hear the warm, energetic air of indoors pour out through this portal into the night sky, where the barometric pressure is lower.

On Lynne's desk, papers flutter like startled pigeons. A man typing at the computer looks up at this sudden whir of noise. "I once clocked a windspeed of ten miles per hour *inside* the tower," he boasts. Then the door slams, shutting off the flow. Papers settle back on the desk. The building has emptied its lungs.

Half an hour passes, and Lynne does not return. By itself, this means nothing; she could be visiting our neighbors across the summit cone at the TV-8 building.

I grab a coat off the rack and head up the metal stairway and out the lower tower door to take an observation. "Doing an ob" is what we call the process. Fog has thickened, and has erased the night sky. According to the red ink on the thermograph, the outside temperature has now risen above freezing to a relatively warm 38 degrees Fahrenheit. The slope of a warm front slides gently over the peak and threatens to turn our thin crust of snow and rime to slush.

"It's the Bahamas," calls out Lynne Host, coming back in through the tower at just that moment.

Already the walls of the building drip with moisture; sheets of rime peel off the tower like wet labels and slide to the ground. A slushy apparition drops off the upper tower and crashes toward my head just outside the doorway; I dodge and weave but get splashed anyway.

The fog intensifies and grows even wetter as the temperature rises. The wind no longer deposits rime. I half-wade through this fog, my arms reaching into the wind as if swimming. With each breath, another drop of mist drips into a cold puddle of moisture at the bottom of my lungs. The fog is so thick now that it blocks the rays of my flashlight like a wall; the beam shoots out ten feet like a bright yellow rod and then shuts off, as if snipped in two.

A trail of puddles—somebody's boot prints, probably Lynne Host's—still marks the remaining crust of ice. The trail meanders into the dark mist and disappears, melting as it goes. I follow the tracks to nowhere. Then I'm forced to retrace my steps.

From the southeast edge of the peak, storm winds suddenly howl with renewed intensity. Gusts seize my shoulders and shove me back toward the tower door.

AFTER MIDNIGHT, THE weather system slides away into Quebec and Maine. The sky, tinted a metallic purple by the first glimmers of the half-moon, clears all in an instant. What happened to all the fog? I wonder. Did some cosmic janitor come along when I wasn't looking and vacuum up the clouds?

Outside, the air is still moist. An occasional sheet of fog materializes on the soft breeze; it snuffs out the stars as if they are distant candles. They reignite a few seconds later.

Fog is an enigmatic presence, the atmosphere turned visible. We see the unseen. I often think we observe *more* details in fog, not less. It opens our eyes rather than blinds us. It paints our world anew, so that even familiar landscapes demand attention lest we stumble or drift off the trail.

"I'd rather hike in fog than without it," an Appalachian Trail segment-hiker explains to me the next morning. It is too late in the season for him to be a through-hiker; the snows will fly before he could possibly reach Katahdin and the end of the trail. The sun bulges on the orange horizon as he speaks. "It's comfortable, like a blinder on a horse. Sometimes on a clear day I'll look up at the summit and it seems too far away, like I'll never get there. I almost give up. With fog you don't have a choice. It's just one foot after the other."

His companion, a bushy-haired nature photographer, sets his camera on a flat boulder and remarks, "Some of my best pictures have been taken in fog."

Distinct species of fog exist, each with its own enchantment. Phrases like "a London fog" and "pea-soup fog" enter the lexicon. When warm rain tumbles from the sky in summer, it saturates the air and forms a dense mat of mist over the ground; weather books call the result "precipitation fog." If a land breeze blows cold air over the oceans, water from the seas evaporates into the dryer air, and suddenly a mist called sea smoke wraps around ships' masts and lighthouses.

"Upslope fog" is the proper and familiar term for the windblown mist now dissipating over the summit. As moist air rises, forced up the mountain by the wind, it cools to the dew point and condenses into cloud. Years ago I witnessed a perfect example of upslope fog at sunrise, just 24 hours after a heat wave had deposited hordes of t-shirt-clad tourists on the summit. I woke at 5 A.M. and saw snow on nearby Mount Clay, a tiny metamorphic ridge just across the Great Gulf.

Snow? Was it possible? After all, it was July and temperatures had been above normal for the season. Oddly enough, the rocks just below my window looked bare—jumbled cones of dry, gray stone. A mostly clear sky framed the scene, an azure ocean in which the sun swam from cloud to cloud. Only Mount Clay appeared white. Surely it could not be snow.

I rubbed by eyes. The illusion did not vanish. What mystery caused that smooth white cloak to be fully draped over Mount Clay? Then the answer hit me. It was a shallow cap of fog, a misty skin pulled taut over the mountain. The cooling wind poured up and over the summit, depositing a wispy film at the top—and only at the top.

I walked up to the weather room and encountered a fellow Observer, Norm Michaels, gazing silently out the window as the first colorful dye of sunlight spread as if by osmosis through the morning air. "What the hell are you doing up so early?" he barked.

Manners at any mountaintop station are somewhat lacking early in the morning, so I wasn't surprised; to wake up properly, most people require coffee and Norm was no exception. In any case, I gave no good excuse for my presence. "I woke up. I thought there was snow on Mount Clay."

My colleague grinned. "I thought the same thing. 'How long did I sleep? Is it winter again already?' But it's forty-five degrees."

What we both saw was liquefied wind. Water vapor that had condensed as a westerly breeze had surged up the far slope and poured over the other side. Rising, the air cooled to the dew point; condensation exceeded evaporation. Fog appeared, white and pure as snow. Then the wind dipped down on the lee side into warmer air, and the mist evaporated. All that was left was a crest of snowy fog, motionless on the barren summit—a white skullcap. Upslope fog is the mist that mountains milk from the sky.

Properly speaking, of course, fog is just a cloud on the ground. Then again, a mountain is nothing but ordinary ground thrust into the realm of clouds. Go to any peak sufficiently high, and the fog and the clouds become one and the same.

18

Going to Extremes

MOUNTAINEERS IGNORE THE SKY'S CHANGING MOODS AT their own peril. "Don't be stupid, because stupid can make you dead," an experienced Himalayan climber once told me. I followed his advice by not going at all.

With due respect to the late Mr. Mallory, "Because it's there" is a poor excuse. There has to be more. The journey to the roof of the world is always one of discovery, exploration, and—especially—risk. What draws people to such a place?

I can only speculate based on my limited experience trudging up and down our tiny Northeastern mountains, plus snippets of conversation with people who have reached much, much higher. Climbers in the Himalayas observe the blue curve of the horizon, dappled with clouds, from a truly unique vantage point. The savage alpine weather, along with physical exertion and anguish, accentuates the beauty of nature, the crisp air, the pastel tint of the rolling landscape. And, of course, there's always the satisfaction of overcoming danger—even if it is simply done by reading a book or gazing awestruck at David Breashears' IMAX depiction of the Khumba Ice Fall at the Boston Museum of Science. The vicarious route is how most people travel to Mount Everest, but it is a popular route nonetheless. Some small piece of us wishes we were there. Or wishes we wished we were there.

Hillary's conquest of Everest was no different from Captain

Robert Falcon Scott's quest for the South Pole (except that Hillary made it back alive). Both undertakings fascinated the public.

Gold is precious because it's rare—difficult to obtain. Is a mountaintop any different? The experience is raw, as are weary climbers' feet. And their emotions. To stand at the apex of the world and watch the whole Earth fall away underfoot, endlessly spinning through space—who wouldn't want to go to such as place, just once?

Me, for one. I admit I want to wish I wanted to go, but that's as far as ambition or sentimentality takes me. Those who do go pay a price. A human being thrust into the thin air of the upper troposphere must learn to endure the true extremes of weather. The human body is woefully under-insulated and ill-equipped. Mountaineers suit up in windproof parkas and synthetic layers of insulation, with gaiters wrapped around their shins and claw-like crampon-points protruding from their toes for traction. A certain percentage of climbers on Mount Everest even choose to supplement their lungs with a mechanical prop: bottled oxygen. The air at 25,000 feet and higher thins until each painful gasp is parched and unsatisfying. You suffocate on a wispy atmosphere that is already two-thirds of the way to being a vacuum; go any higher and you should have brought a spacesuit instead of an ice ax.

The dryness of the air on the world's highest mountains is pernicious; it rakes at your lungs. Hacking coughs ensue and refuse to abate. The fact that the air is so attenuated, with proportionally less oxygen, nitrogen, and moisture, makes it all the less pleasant to inhale. Where one breath satisfies you at sea level, you must gulp down three or four on Mount Everest. Your body loses more moisture to respiration than to perspiration; dehydration is a danger. Blood thickens to a reddish sludge. The number of red blood cells in your arteries doubles to provide more oxygen to suffocating cells and organs. Capillaries in your eyes exert pressure outward, only to find little inward atmospheric pressure to balance the push; they hemorrhage. Your bloodshot eyes possess the ochre-red hue of storm clouds at sunrise.

At high altitude, the atmosphere is alien, vaporous, and bitter cold. How alien? Frequently, temperatures along the equatorial regions of the planet Mars will warm until they are downright hospitable compared to the pinnacle of Mount Everest on an icy night in May. Neither Mars nor Everest is a comfortable locale for the unprepared—but to which place would you rather walk?

From a meteorological point of view, Mount Everest hoists a pillar of rock almost to the roof of the troposphere. It neatly dissects the lowest layer of our atmosphere, where nearly all weather occurs. In this way, the mountain provides a convenient ramp upon which climbers may view the stratification of Earth's air. Just as the layered rocks of the Grand Canyon yield a geologic cross section of millions of years of deep time, mountains like Everest give weather watchers easy access to the sky. To hoist yourself from the syrupy, nourishing air at sea level to the noisy, rarefied current of the jet stream, you don't need wings. You only need the strength of will—and the luck—to climb the high Himalaya and come back alive.

Not everyone succeeds. Lady luck abandons the unwary; fair weather turns foul. Approximately one out of every ten climbers who have set foot on Everest's upper slopes has returned to base camp in a body bag—or not at all. "Sometimes you see a dead body, just left there. They never made it," I overheard a climber reminisce.

I cannot claim to have trudged very high on foot. (Trips to the jet stream in commercial airliners don't count.) I lack the sinews, the lung power, or the recklessness to try. In my experience, the weather of lesser mountains here in New England can turn savage enough.

"MOUNTAINS ARE WEATHER makers," says a man on the trail behind me, echoing my thoughts. I don't mean to eavesdrop, but the crisp wind carries his voice clearly for a quarter-mile.

I am climbing to the observatory at what seems to me a fast pace—but apparently not fast enough. Looking back, I see the man leading a party of six hikers, all of them steadily overtaking me near the top of the Tuckerman Ravine trail. In single file they scramble uphill, each gripping a pair of multicolored ski poles for support and balance. The rhythmic motion of their arms and poles as they piston forward makes the whole procession look like a steam locomotive; the smoky condensation of their breath in the early-morning air adds to the illusion.

Phrases and random words from their conversation ripple through the air as sound waves, then translate back to meaningful language in my ears and brain. Based on what I hear, the group's leader is a Himalayan veteran, slumming in the Appalachians on a late summer day. He tackles the Tuckerman trail like an energetic child sprinting up a staircase two steps at a time. The group is now little more than 100 feet behind me. In answer to an unheard question, the leader utters a single succinct sentence. "Severe weather," he intones, "is weather which cannot be ignored."

I've heard that before—it's a good motto for climbing mountains. The man speaking those words is tall, clad in black windpants with a blue parka that billows in the breeze. He steps off the trail on the summit cone just short of the geographic peak, raises a hand to his eyebrows, and gazes west across New Hampshire into the distant azure ridges of Vermont's Green Mountains. The prevailing visibility tops 70 miles. The man's shadow, like that of the mountain we stand on, slants west across two neighboring states. The ruddy sun glimmers coldly far below his feet, newly risen. For uncounted seconds, the sun stays anchored to the horizon.

The morning air is deliciously thin and crisp. Overhead, the sky stretches forever in a smooth blue curve. I faintly detect the fibrous line of cirrus clouds advancing from the southwest. A thicker, lower layer of stratocumulus ripples and slides over the nearby green valleys. Distant peaks protrude above the clouds in the north.

Suddenly, the mountain breeze ceases to murmur; it bites and snarls with the chill of September. The guide's hair—what little he has left, a clump at the back of his skull—is as white and bright as the thin crust of rime on the boulders. We are all standing on the highest visible pinnacle of land, and the temperature is just starting to top the freezing mark. But the man is not wearing a hat; his ears whiten, nipped by frost, numbed by cold.

Without speaking, I walk past the guide and his team and step into the safety of the summit building. I am ready to shed this 40-pound backpack and put up my feet.

An unexpected gust of warm air slams me in the face as I open the door. Artificially heated air inside the building is always swirling with energy, and today it is nearly saturated with moisture from the breath of hikers and park rangers. Contained by walls and windows, the indoor air tries to expand against the boundaries; it waits for the seal of the door to open, then surges through the crack with an explosive, invisible current. It dissipates unseen in the cold outdoors.

ONLY THE TALLEST mountains on Earth— the Himalayas of Nepal, Tibet, India, and Pakistan—intrude into the sky as high as the river of wind we call the jet stream. These icy platforms of land are constantly pummeled by whips of wind.

Jon Krakauer's 1997 mountaineering epic, *Into Thin Air,* met with an avalanche of success, an indication that people who would never dream of visiting Mount Everest in person still want to know what it's like. Usually, men and women who have seen the summit are happy to oblige the curious who have not. "When it's dark out, you're in your tent. If you're out at night, you're really screwed up," announces a middle-aged man named Joe, born with the compact, wiry build common to climbers but compensated with a confidence bordering on arrogance. He is sitting in the kitchen of the observatory, swapping tales with a crew of meteorologists. He describes this scene: at a high camp in the Himalayas years ago, he listened to the jet stream shriek all night against the tallest wall of rock in the world. For a while he was awed, shivering. "The thin air makes it harder to keep warm," he explains. Direct sunlight provides an illusion of warmth during the daylight hours, but the scarcity of air molecules at elevation keeps the temperature low.

I listen with half an ear, fixing a snack in the pantry to take upstairs to the weather room; I am still on duty. A second visitor, his arms knotted with muscle, brings his own story to the table. I don't catch his name. "In 1995, seven climbers were trapped above eight thousand meters on K2, a really nasty mountain." Pictures of K2 support his view—it is a jagged tooth of rock, a formidable technical climb compared to Everest, and almost as tall. One out of every three climbers who challenge K2 never returns. Despite the odds, many try.

The man at the kitchen table tells his dumbstruck circle of listeners how, through binoculars, viewers at base camp might have watched the wind pluck victims one by one off a high ledge and fling them directly to their deaths.

I don't hear the rest of the story. Work draws me away; I have a weather observation to take and file, so I hurry upstairs. Outside, the wind scarcely whispers. I swing a thermometer and wait for the wet bulb to drop. A warm, friendly sun bakes the ground. How tame, I think, compared to the saga I had just overheard.

Weeks later, however, I looked up the man's story and began to suspect that it had been Paul Bunyanized, exaggerated, the way all epics of conquest and tragedy are with the inevitable passage of time. In this case, though, the details were distorted only to a slight degree.

Newspapers and magazines tell the tale in equally dramatic (and, sometimes, equally exaggerated) fashion. On August 13, 1995, two Spanish climbers retreated from the wind at Camp Four on K2 after savage gusts ripped apart their tents. During the long trudge downhill, they discovered a clump of gear in the snow—the boots, anorak, and harness of 33-year-old Scottish mountaineer Alison Hargreaves. Farther down the trail, near Camp Three, they located evidence in the snow and ice indicating that at least three climbers had fallen from the summit ridge, 4,000 feet overhead.

Seven people died on the world's second-highest mountain that night. It was both ironic and tragic, because the day had started in triumph. Hargreaves and her party had stood on the 28,250-foot summit and informed base camp of their victory with gleeful voices. Then fate took a wrong turn. Winds gusted to hurricane-force or greater. It is now believed that only a short time later, Alison Hargreaves and her companions were ripped loose from the summit ridge by wind and hurled to their deaths. Their final radio communication with base

camp celebrated the success of the ascent. And then the wind silenced those voices forever.

DEATH IN THE mountains often wears a white mask, a sheet of
snow. People who perish in avalanches are suffocated by a scarf of
solid snow. The victims' own warm breath melts the snow as they push
and paw frantically inside their icy coffins. Desperate to claw a way
out, they toss and turn, scrape and dig. But the air never lasts—slowly
each victim suffocates inside a white tomb. Any snow that melted due
to the heat of their last panicked breaths soon freezes solid—a death
mask. The victims lapse into unconsciousness as the wet ice stiffens.

"If you're not rescued in the first hour, chances are you're dead,"
a rescue worker in Tuckerman Ravine explains to me. She is a short,
plucky woman who folds her arms and nods to punctuate that last
word—"dead." Each avalanche is a snowy surge of sudden violence
that requires a quick rescue. The mountain gropes with a white paw
and swats stray hikers off its hide.

For mountaineers, hypothermia is just as deadly. But hypothermia is a stealthy, cunning hunter, slow to creep up on its prey. This
killer is too subtle. Sometimes, when it already has you tight in its
jaws, you forget you are dying.

"What hypothermia does first is steal your brain," adds another
veteran of search-and-rescue operations, a man with an untidy brown
beard like a tangle of bristles and briars. "It doesn't matter how well
equipped you are. You could have ten sleeping bags, a tent, and a cellular phone, and you'd just stare at them and slowly freeze to death."

It is late afternoon, and the three of us and at least 20 others
meet near the glacial tarns named Lakes of the Clouds on the slopes
below Mount Washington. We have been called to the scene for a rescue that fortunately involves nothing more serious than a twisted an-
kle. But the wind buffets us at 70 mph, so the three-hour litter carry
lasts longer than expected. Temperatures hover in the 50s and the air
is dry. Two dozen rescuers take turns either lifting the dead weight of

202 • E X T R E M E   M E A S U R E S

the injured hiker in the litter or else resting by walking to one side, out
of earshot of the rescuee. There, the talk predictably turns to other accidents, other experiences. Relatively few have happy endings.

I remember that one winter, after two days of fierce wind, a death
by hypothermia was noted in the observatory logbook: "A fine day on
the Rockpile, as everyone gets out for some fresh air and sunshine.
Winds died down to a wimpy 24 mph by early afternoon. But a lost
hiker was found dead on Mount Eisenhower shortly after 6 P.M. Not a
good day for him, or for the rescuers."

By far the deadliest mountain in the world is Everest; more than
140 people have perished there. The causes vary: the cold grip of the
wind, slips and falls, or the inhalation of rarefied air that addles
people's brains—or sometimes a combination of them all.

The litany of disasters that can overtake mountaineers reads like
a roll call of weather-related phenomena: frostbite, avalanche, hypothermia, hurricane-force wind, blowing snow, freezing fog. In the
well-publicized tragedy on Everest that took five lives in 1996, including those of experienced guides Rob Hall and Scott Fischer, the lack
of oxygen surely took its toll. When one unlucky climber seated himself in his own grave in the snow and simply waited for the end, no
rescue party was able to arrive in time to save him. The lack of oxygen
probably deprived him of the full significance of that fact.

K2, and Denali in Alaska, also possess reputations for danger and
a long tally of victims. But surprisingly, not far down the list of fatal
mountains is a tiny New England hillock named after the first President of the United States—Mount Washington. More than 125
people who did not take the danger seriously paid for that transgres-
sion with their lives.

The name or size of a mountain does not really matter. If you
make a foolish mistake while hiking or climbing, you are likely to pay
in full. The alpine environment does not make change.

"When I came back I was twelve pounds lighter and a lot
weaker," announces a tall, thin man named Ken Pickren. I meet him
during an EduTrip on top of Mount Washington, where he talks ea-

gerly about his recent trek to 22,834-foot Mount Aconcagua, in Argentina. At that altitude, high in the troposphere, more than half the atmosphere is already below you. "You do a lot of useless breathing," Pickren explains. "You do a technique called pursed-lips breathing." He demonstrates for a second or two, huffing and puffing. "It offsets the partial pressure in your lungs." How it accomplishes that feat is by opening up the alveoli, bringing oxygen more easily to the bloodstream. It makes the air feel more like low altitude. "You have to concentrate to do it, though, and you often don't think about it 'cause your brain isn't working," says Pickren.

On the way up to the summit, Pickren and a friend camped at 13,000 feet to acclimatize. In the thin air, night's chill sets in quickly. "As soon as the sun went down you'd reach for your fleece and get in your tent. At night it would freeze up on you. You'd have frost in your tent." The temperature plummeted into the 30s Fahrenheit. "In the morning, standing in direct sunlight, it was maybe fifty, sixty," he says.

Wind howled at a constant 40 to 50 mph; Pickren comments on how steady it was, lacking gusts or lulls. "One guy, he lost his stuff-sack cause he put it down and it was gone." He waves his hand dramatically. "It's probably in Chile."

One morning he heard a stir in a neighboring tent. A woman was talking loudly, trying to rouse her husband from an unusually stubborn slumber. Turned out the man had cerebral edema, altitude sickness. Other hikers rushed to help. "He was hallucinating, flailing his arms," says Pickren. "It took eleven of us to hold him down. He was physically abusive." Quickly, they placed the sick man in a special medical bag which mimicked the air pressure of lower elevations. That brought him round again. "In fifteen minutes he was lucid. We took him off the mountain as fast as we could."

Pickren's adventures weren't over yet. He still had a summit to climb. "We were in one nasty storm. We got up one morning and there was three feet of snow." In general, he says, it snowed every morning.

Finally he stood at 22,700 feet, just a 135-foot vertical climb from the ultimate goal. The wind roared. "It sounded like ten jet engines right over my head. The winds increase by an order of magnitude as you approach the summit." And so he did what many climbers do not do when the weather turns foul but the summit is tantalizingly close: he turned back. That decision may have saved his life.

"I did not reach the summit. I have no illusions about that," Pickren says without apology. He quotes the famous mountaineer Ed Viesturs to defend his choice: "I had a choice of coming home or staying there." Forever.

DURING THE FILMING of brief portions of the IMAX Everest movie in New Hampshire in 1997, observatory crew member Dar Gibson offered to lend a hand. "I was a double for Tenzing Norgay's son at one point," he remembers. The total footage from Mount Washington amounts to less than 90 seconds in the actual movie, but it took a full week to shoot. The weather cooperated, providing alpine scenery and conditions severe enough to resemble a high camp on Everest, but much more accessible. "It was probably twenty below zero. Pretty much it was windy and cold," says Gibson, an avid snowboarder and climber in his mid-20s. He rubs a hand through his short-cropped hair, thinking back.

For one scene, the yellow Mountain Hardware tents were filmed straining in the hurricane-force winds. Gibson was inside one tent, holding on for dear life. "I was spread-eagled in the tent with a couple of backpacks. I was ballast, holding it down. But the wind was getting underneath the tent, threatening to roll the whole setup right down the cliff." Outside, the film crew wrestled with the camera. Gibson recalls being shaken and tossed by the savage gusts. "I was pretty scared. It felt like I was a set of dice in a cup."

The whole ordeal lasted about approximately 40 minutes. "If I just sat on the tent floor, parts of it lifted up. I could hear them yelling and trying to talk to each other outside the tent, and I was trying to yell back at them, but we couldn't hear anything."

The filming on Mount Washington was mostly filler: background scenes, tents flapping in the breeze, climbers walking up a slope with their faces hidden. "After it got dark one night, they had us put on headlamps and walk slowly toward the camera as if we were just trudging along at 28,000 feet or so, at altitude," Gibson says. With the relatively thick air and plentiful oxygen on Mount Washington, the instinct was to step quickly. They had to force themselves to slow down to portray the slow-motion pace of a climber hindered by hypoxia.

"I was sort of a summit guide for the film crew. They didn't always know where they were going with the fog and blowing snow," says Gibson. Not all of the film crew were experienced climbers or had been to Everest. "I was worried about some of the film guys. Windchill was super-low, and I wasn't sure they were prepared. After a while I said we better go in—and they were ready to go."

DAR GIBSON LIVED and worked for a year on the summit of Mount Washington. After his largely anonymous 15 seconds of fame in a movie about Mount Everest (I asked him if he was credited as "the guy in the tent"), he moved on. In 1998, he followed in the footsteps of a dozen other Mount Washington Observers, heading for a point where the longitude lines—and, in a way, the winds of planet Earth— converge. Where the weather reaches its coldest levels. Antarctica.

19

Chasing the Sun
from Pole to Pole

WHEN THE EXPLORER ROBERT FALCON SCOTT SAILED HIS ship *Discovery* into the often-stormy band of water below 60 degrees south latitude in November 1901, the wind cast an icy spray onto his newly grown beard. Never before had Scott visited the polar regions, and according to his journal, the first sight of icebergs and sea ice awed him. "What light remained was reflected in a ghostly glimmer from the white surface of the pack," he wrote.

Scott and his scientific crew of 123 men made landfall at Ross Island at 78 degrees south latitude in February 1902. Scott was a lieutenant in the Royal Navy, and his mission was to measure the size of the Antarctic landmass and determine the depth of the ice which covered it.

The expedition prepared to buckle down for the winter. Aboard ship, they possessed two years' worth of supplies, as well as pre-cut timber to construct shelter. Camped on the edge of an ice shelf the size of France—the Ross Ice Shelf—with a view of volcanic Mount Erebus smoking in the distance, Scott and his men were only the third group of human beings ever to winter over on the coast of the southern continent—and only the second who actually intended to do so.

The first winter encampment in the Antarctic was purely an accident. It had occurred just a few years earlier, in 1898, when the swift onset of the cold season forced a Belgian named Adrien de Gerlache to prolong his visit after sea ice trapped his ship, the *Belgica*. (Each winter, the freezing of sea ice around the coast effectively doubles the size of Antarctica, to nearly 30 million square miles.) From de Gerlache's point of view, that meant trouble; by the end of the winter, he probably wished he'd stowed away a few extra blankets. The following summer, he and his crew frantically chopped their ship free. Their ordeal is a very minor footnote to history, and I mention it only because, coincidentally, among de Gerlache's staff was a young Roald Amundsen, first mate—the same man who would later become Scott's chief rival in a deadly race to the South Pole.*

A race to the pole was not yet in Scott's mind in 1902 at the start of his first winter in Antarctica. In an area now known as McMurdo Sound, he and his crew melted ice for water and braced against the constantly raging storms inside their timber shacks. Soon the sun set for the last time, and they were truly alone. The air thickened and cooled; the ocean froze. Sea ice encased the *Discovery* until 1904.

"I ONCE WATCHED a sunset for five days," says Mark Ross-Parent, three years after his final stint on the world's coldest and least populated continent.

The sun, of course, is what powers weather on Earth. Where better to study certain facets of meteorology than in a land upon which the sun shines continuously for up to six months, only to vanish altogether for the next half of the year? Given the curious behavior of the sun at the bottom of the world—and the resulting savagery of weather—it should surprise no one to learn that the careers of more than a dozen meteorologists have drifted from Mount Washington to frigid Antarctica and back again.

In Antarctica, Mark recalls, the autumn sun circles the horizon day after day, like an indecisive eagle searching for a place to land. Jet-black night lingers for nearly five months at the pole, with an eerie twi-

---

In 1899, the first "planned" winter encampment on the Antarctic mainland was led by an English and Norwegian adventurer named Carsten Borchgrevink.

light that endures for weeks at both ends. It's like living on a different planet.

At the moment, Mark is half a world away and at a place 160 degrees warmer than the South Pole. He is standing in a vegetable garden in the northeast corner of the United States on a sunny August morning. The shoulder of his t-shirt is streaked with mud. The air here is sticky and humid, thanks to a warm front pushing north from the Carolinas. Cotton-ball cumulus clouds blossom in the sky underneath a thickening layer of altostratus. The red column of fluid in a nearby thermometer reads 80 degrees Fahrenheit. But Mark clearly remembers the chill of Antarctic winters. Even now, as summer breezes heat New Hampshire and the jet stream swerves north into Canada, the depth of winter darkens the bottom of the world.

For weeks in March on the cusp of winter, sunlight trickles away, dissolving slowly, not to return for half an age. Twilight lingers. Then night clamps down firmly and does not let go. For months on the ice-capped desert at the bottom of the world, the breeze is set in motion by a sun it cannot see.

Mark assures me that anybody who has ever lived through a six-month winter night at the pole feels that light is the vital missing ingredient. He recalls one bitterly cold night at the pole in which the brightest light was manmade. It was during a midwinter airdrop, says Mark, that the South Pole crew set out "smudge pots," five-gallon barrels filled with gasoline.

"You torch them," Mark says, referring to the smudge pots. "We set up a large rectangle the size of a yard. It was the coldest day of the winter, a hundred and eight below zero. You didn't have to get too far away from the fire before you could feel the frigid cold. The cold dominated the bonfire. In fact, I think the snow and ice only melted down about a foot."

The fires provided a beacon so that an overflying pilot would know where to drop the mail. The mail (or any contact with the outside world) is just as treasured as light during the prolonged Antarctic nighttime. It is too cold to land a plane in winter at the South Pole—temperatures below minus-60 stretch the safety limits—so a quick fly-by is the best they can do. (The custom of the midwinter airdrop actually stopped in the mid-1990s—too dangerous for the pilot and too disappointing for the crew members who don't get mail.) Then the fires burn out and the South Pole crew is alone once more.

Inside his home in Jackson, New Hampshire, Mark shows me a picture of the fire, a blazing inferno that scars the night sky. It's hard to imagine feeling winter's chill so near to a flame tall enough to engulf a two-story house. I have never experienced an ambient temperature lower than minus-40—and that's like a balmy winter night at the pole.

Smudge pots and mail-drop beacons are by no means the only wintertime radiance visible from the pole. High pressure and clear skies dominate the weather forecasts, so starlight often glimmers through the rarefied air. The moon looms like a giant silver disk just overhead. The most noticeable light is the aurora australis, or southern lights, which frequently blaze and sparkle in waves of ionized light far overhead.

The aurora is a cold, distant flame. Ionized molecules float high in Earth's atmosphere, creating red and green ribbons that dance across the sky at an altitude of 50 to 70 miles or higher. It is one of the few easily recognizable "weather" phenomena occurring above the troposphere. Molecules ingest the sun's rays, energize, and then ignite colorfully as they return to their normal, lesser state of energy. Nitrogen turns red or violet, while oxygen glows red and green.

The source of southern (and northern) lights, like all weather events, is the sun. Protons and other tiny particles are hurled outward from our local star on the solar wind, creating the aurora; these particles intercept Earth's air far above the troposphere, flying through the solar system at 200 to 600 miles per second. The planet's magnetic field funnels these sheets of color and energy toward the poles. For that reason, dramatic auroral displays in the tropics are rare. But strong solar flares enhance the show and sometimes allow it to be seen at latitudes close to the equator.

I once heard the aurora described poetically as "where the sun's atmosphere touches our own."

"In August, it's still basically dark down there," comments an Antarctic veteran, Navy meteorologist Jack Halpin. He recalls a spectacular display of aurora he witnessed from the coast at McMurdo Station. "It was a real weird night. We had southern lights, a lunar eclipse, and a meteor shower going on at the same time."

"Sometimes they would last minutes, other times they would last days," Mark Ross-Parent confirms. His 35-millimeter slides of the aurora show whirls of green and crimson 60 miles high. "They were generally greenish or whitish in color, occasionally red or pink. Sometimes blue. Whether it was a good year or not, I don't know. But it was the most I've ever seen."

DURING HIS TIME "on the ice," Mark once worked for a season at the now-defunct Siple Station, named after the inventor of the wind-chill chart, Paul Siple. "It's a pretty remote station. It's not in existence now. It's closed. I was there for the last season."

The primary function of Siple Station was to monitor the ionosphere, that high region of the atmosphere approximately 30 to 250 miles above the surface, at which altitude the surge of the solar wind violently strips atoms of their electrons and ignites the aurora. The ionosphere overlaps the mesosphere and the thermosphere. It is loosely defined as the part of the atmosphere where enough ions and free electrons exist to effect the transmission of radio waves—reflecting radio signals back to Earth, so that listeners in Maryland and Virginia can hear broadcasts from faraway stations like KMOX in St. Louis, Missouri.

The NASA space shuttle flies through the ionosphere, but the air is so thin at that height that the atmosphere is scarcely different from outer space. Instruments at Siple "were primarily used to study low-frequency radio waves. They had huge antennas, miles long," Mark tells me. He lived and worked there in true isolation, with fewer than ten companions.

In 1945, Paul Siple, the station's namesake, invented a formula for calculating the numbing effect that wind has on the human body—we call it windchill. Meteorologists will define windchill as "how cold the air feels, not how cold it actually *is*." The bloodstream and circulatory system of human beings always emanate body heat, which forms a thin, invisible shell of heated air that clings to our skin. It is a layer of subtle, invisible insulation less than a centimeter in width. As I mentioned earlier, even a soft breeze can skim this layer away; if the wind increases enough, the protective shell will disappear altogether. The faster the wind blows, the quicker our body heat escapes through this rapidly thinning shield of warm air. That's windchill in a nutshell.

Most windchill charts stop at 40 or 50 mph, with the additional disclaimer, "Wind speeds above fifty miles per hour provide little additional effect." I suppose it scarcely matters to most of us if we get frostbitten in one second or two or three. Either way, the solidification of skin cells hurts.

Still, I've often wondered if windchill really does stop getting worse after 50 mph. In Siple's original mathematical formula it does— the equation even causes the calculated temperature to *rise* after a certain point. Of course, it's not really feeling warmer as gusts increase from 50 mph to 80. Experience and common sense dictate otherwise. It's just a mathematical fluke. But how can we prove it?

About a decade ago, a technician from Mount Washington decided to test the windchill principal on cold winter days. Winds gusted to hurricane-force. Few places on Earth ever need a windchill chart above 50 mph, I suppose, so the quirkiness of Siple's formula is rarely an issue. But on Mount Washington, the strong winds demand attention. And at least one Observer was happy to oblige.

With a stopwatch, a technician named Greg Gordon exposed his bare hand to the wind, timing to the exact second how long it took for the first whitening patch of minor frostbite to appear on his arm. (This is true.) He attempted this feat multiple times, in slower winds, faster winds, and changing temperatures. In the end, he even plotted the data on a graph and wrote up a tongue-in-cheek report in the Mount Washington Observatory's news bulletin. The conclusion: winds above 50 mph do indeed have additional effect.

Clearly this man had too much time—and frostbite—on his hands. "But that's what I call a dedication to research," said one of his colleagues years later, laughing.

THERE ARE CERTAIN ambitious things I've always wished to do—read *Moby-Dick* all the way through, take up hang-gliding, et cetera—and a visit to Antarctica is one of them. It's not too late, of course. But listening to the tales of colleagues makes me wish I had already gone myself.

I learn I'm not the only one who feels that way. "I think everyone who comes through here has wanted to go to Antarctica at some point," says Sarah Curtis, sitting in the weather office at the observatory. She laughs. "I actually think working here for two winters got it out of my system." What fired up her imagination were the stories of Mark Ross-Parent, her supervisor and trainer in 1996, when she first arrived at the observatory as a meteorology undergraduate. "That's what excited me—that if I put in the effort, I could do that if I wanted to. And for a while, that *is* what I wanted to do. But it's so hard being away from your loved ones every other week here at the observatory, much less a thirteen-month stint at the pole." The enthusiasm in her voice wanes. "I think that at this point, being with those people is more important. If I was younger, or if I was not ready to get married, I think I would do it. I think people need to keep reevaluating what's important to them at the moment, whether it's with their career or with their family. And while Antarctica seemed like a great idea at the time, a lot of things change."

The bottom of the world is not for everyone; that much was clear from the start. Antarctica is a land so cold that ordinary mercury thermometers freeze solid, and the sun is at times a distant rumor. And yet, for many, the extreme climate is the sole reason that they come.

Around the kitchen table at the observatory, I once discussed Antarctica with Navy meteorologist Jack Halpin. We talked about the peculiar winds that pour down off the ice and tumble to the sea. Workers in Antarctica call them "Herbies."

"Herbies" are downslope winds, similar in origin to the descending torrents of wind called Chinooks in the warmer climes of the North American Rockies. "They're actually density winds, katabatic winds," explains Halpin. Katabatic winds start as inversion winds—masses of dense air pulled downslope by gravity. Over time, the winds are channeled by the surrounding landscape—mountains, for example—finally converging to produce a volatile scream of air. The windiest place on Earth is probably Antarctica's Adélie Land, in the region of Cape Denison and Commonwealth Bay, where the average annual wind speed tops 50 mph (80 kph). Adélie Land is adjacent to the ocean, and winds that began high on the East Antarctic Ice Sheet are here at their swiftest and loudest following a 900-mile downhill plunge.

Jack Halpin saw his share of katabatic winds when he worked at McMurdo Station in 1978 and 1979. Halpin is a middle-aged man with a mop of brown-gray hair and a rich trove of stories from his Navy years, which he eagerly relays in a tangy Maine accent to anyone willing to listen. We swap anecdotes for a while across the kitchen table. I ask about the meteorological origins of Herbies. To illustrate, Jack reaches for a book the size of a small desktop—a world atlas—and flips through the index.

"High pressure builds over East Antarctica," he says, pointing to the center of the sprawling atlas, which is now open to a map of the southern continent. "The air's very cold and very dense and it spills down off the continent and will come through some of these glacial areas." His finger traces a line across the map to Beardmore Glacier. McMurdo Station is stamped in black ink at the edge of the sea. "We could actually see the storm like a sandstorm, except it was snow in the air coming our way."

A week earlier, when I had spoken to Mark Ross-Parent in his vegetable garden in Jackson, he too described the experience of wading neck-deep through the forceful blast of a Herbie. "We could literally see a wall of white. You could see the storm coming."

When I later asked both scientists how "Herbies" got their name, neither knew the answer. It's a bit of a mystery. "That's what we called them, Herbies. But I don't know why," Mark told me. "I haven't heard that term for years."

"No one even has a guess on it," added Jack seven days later.

Unlike a typical low pressure storm in the United States, a Herbie is created by high pressure—sinking air. Colder, denser air rolls down the slopes, gathering speed as it goes. It's a runaway avalanche of oxygen and nitrogen. "Just like water rolling downhill," Dave Thurlow comments. "Antarctica just slopes to sea level from the center of the continent on down. It can blow steady at one hundred miles per hour for days and days."

"We'd get hurricane-force winds, sixty to eighty mph and higher gusts," Jack Halpin remembers. "They'd pulsate, too. We'd see jumps in the microbarograph. It was rising all the time."

Back in the observatory, Jack moves his hands over the atlas and describes the effects of a typical low-pressure storm system spiraling across the ocean near McMurdo Sound. The location of a low so close to Antarctica's powerful high can add ferociously to the strength and speed of any Herbie. "What would complicate things sometimes was if we had low pressure in the Ross Sea. That would speed up the wind pouring off East Antarctica, and bring clouds and more snow. So there were actually two types of Herbies." His hand traces an imaginary storm track around the coast.

On a weather map, whenever a high and a low perch in roughly the same location, isobars press together like the lines on a topographical depiction of a steep hill. Each isobar represents a specific atmospheric pressure; stay on the line, and the pressure remains the same. But as wind floods swiftly from high pressure to low, the sky shrieks and screams.

Storms at sea are all too common in the southern hemisphere. Between 40 and 60 degrees south latitude, nothing exists but flat ocean and a few scattered islands; the storms rage unchecked. "There's no continents to stop them," says Jack Halpin, waving his hand across a vast blue expanse on the bottom half of the atlas. At these latitudes, as in the northern hemisphere, a Ferrel cell serves up the prevailing westerlies. A semipermanent low pressure zone called the Antarctic Circumpolar Trough surrounds the southernmost continent from 60 to 65 degrees south latitude. Moist air from middle latitudes seeps toward the pole in this region, producing cloudy, gale-prone weather. The storms in turn push a powerful, looping ocean current called the West Wind Drift all the way around the globe. Even closer to the continent, in a narrow band between the coast and the Antarctic Circumpolar Trough, the Polar Easterlies reign. As katabatic and inversion winds flow off the continent, they veer in a westward direction under the influence of the Coriolis effect. (A wind is considered easterly if it is blowing *from* the east *to* the west. A west wind blows from west to east.) The net result of all these interlocking meteorological gyres is often savage, unstoppable weather. "Sometimes the storms would last a week, and that would stop all the airplanes for that week," says Halpin. "The low-pressure systems are much bigger down there, because there isn't any land. They get enormous."

THE SIZE AND ferocity of the tempests in southern seas delayed the discovery of Antarctica for many centuries. Robert Falcon Scott and his crew no doubt listened to these storms many times as they huddled in their bunks during that first winter encampment in 1902. The success of their mission made Antarctica the final land ever settled by human beings; no compelling evidence exists of any long-term human settlements on Antarctica before the twentieth century. In fact, no human eyes (that we know of) ever gazed upon the coldest continent until as late as 1820.

The mere existence of a vast southern landmass sparked great debate in scientific circles for thousands of years. The ancient Greek philosophers steadfastly believed that a land called Antichthon or Antarktikos lay somewhere far south of all known lands—a twin to the cold Arctic regions that European sailors had already discovered in the far north. Because snowy, jagged islands were already known to exist under the constellation Arktos, the Bear—commonly known today as the Big Dipper—the ancient philosophers argued that similar lands must also exist in the south. "Nature demands symmetry," they declared. They already knew about Scandinavia, Siberia, Scotland, and Iceland; surely equally large areas of land must exist in the south, to provide balance.

Aristotle himself deemed the idea logical, and that was enough to keep the concept of Antarctica alive and in circulation for 2,000 years, despite the absence of any evidence or eyewitnesses.

Who or what lived on this mysterious southern landmass was anybody's guess. A few imaginative European sailors believed that any such land—if it truly existed at all—must be totally inaccessible. Any ship's captain worth his salt knew it was impossible to cross the equator, where the sun glared down from the zenith and caused ships to catch fire.

In the fourteenth century, navigators from the great sailing nations of Portugal and Spain started to speculate that the northernmost expanse of this mystical southern continent was Africa itself. Vasco da Gama shattered that belief in 1497 when he rounded the Cape of Good Hope and sailed into the warm waters of the Indian Ocean. (He also crossed the equator, thoroughly demolishing another myth.)

Years later, Ferdinand Magellan and his fleet sailed around the tip of South America. But Magellan made a mistake; he erroneously marked Tierra del Fuego as the northern extension of the mysterious land known as Terra Australis Nondam Cognita (Antarctica). Maps in the 1400s and later centuries featured the southern continent prominently, though it still had not been seen or explored.

By the late sixteenth century, a majority of mapmakers depicted "Antarctica" as a landmass bigger than North and South America combined—a vast bulge of solid ground sprawling from the pole almost to the equator. Clearly, a hefty prize awaited whomever found it first. Queen Elizabeth and her fellow European monarchs sponsored expeditions to claim and exploit this promising but elusive new territory. They could not afford to let such a treasure of natural resources slip into the hands of enemies.

Little did they know that only ice and wind awaited them. The anticipated paradise of gold and tropical fruits never materialized, except in the minds of dreamy cartographers.

Near the end of the sixteenth century, Sir Francis Drake drifted into the windy, landless stretch of ocean south of 40 degrees latitude. Here, the sinking air from the subtropical doldrums at 30 degrees south is nudged to the left by the Coriolis effect; it spreads into the prevailing westerlies and howls unchecked across the sea. Drake saw nothing but waves and wind, and reported his failure to Queen Elizabeth.

Belief in Antarctica began to wane after Drake's mission. The hope of discovering this new land all but dissolved when Captain James Cook sailed as far as 71 degrees south latitude in 1775 and witnessed only a barren ice pack and towering icebergs. Even if there were land to locate south of that point, what use could it be? At Cook's longitude—106 degrees, 54 minutes west—he was still hundreds of miles away from striking continental land. "No continent was to be found in this ocean but must be so far south as to be wholly inaccessible on account of ice," he declared.

On maps, the mystical continent shrank in size. Surely it couldn't be very big if no one could find it.

The quest for Antarctica finally ended on January 29, 1820, when British citizens Edward Bransfield and William Smith claimed to see a rugged land at more than 62 degrees south latitude. (Their longitude was different from Cook's.) But now a new dispute arose. The possibility that Bransfield and Smith had sighted merely one of the Shetland Islands opened the door for 20-year-old Connecticut skipper Nathaniel Brown Palmer to claim first sighting of Antarctica in November 1820. His sloop *Hero* joined a seal-hunting fleet and sighted what appeared to be continental land on November 20. Interestingly enough, Russia's Thaddeus Bellingshausen laid claim to the discovery of Antarctica in the same year.

For decades thereafter, whale- and seal-hunting fleets represented the chief interest for humans in Antarctica. If no gold or riches existed there, at least there were blubber, whale oil, and furs. The morality of killing intelligent mammals did not yet enter into the equation; whales were still thought of as "fish." Several nations also made territorial claims to sections of Antarctica in the nineteenth and early twentieth centuries, as wars between members of our own species were fought on a massive scale, often over bits of religious trivia or the rights to lands and resources.

The Antarctic Treaty of 1961 stopped all that, at least on one continent. The Treaty set aside Antarctica for peaceful purposes, primarily scientific ones. Sovereign nations which had already made claims elected not to enforce them for the duration of the Treaty. Alone among lands on Earth, Antarctica is effectively owned by no one.

The Treaty preserves Robert Falcon Scott's original outpost, Hut Point, as an historical landmark. Furthermore, according to the Treaty, "Antarctica shall continue for ever to be used exclusively for peaceful purposes and shall not become the scene or object of international discord." Scott came to the Antarctic under the banner of exploration, not for military purposes, and the Treaty attempts to preserve that attitude. Forty-three signatory nations have agreed to put on hold their rivalries and territorial claims for the duration of the Treaty; today, the icy land is mostly used for research. As one textbook title puts it, Antarctica is "a continent for science."

The landscape of Antarctica today is little different than it was in Scott's era, and the population is only slightly larger. At the metropolis of McMurdo Station, or "Mac Town," not far from where Scott positioned his own Hut Point, the maximum summer population reaches about 1,000 to 1,400 people. That number drops to less than 200 during the winter. At the Amundsen-Scott South Pole Station, the numbers drop even more. "Population of South Pole this winter is 41. 32 men, 9 women," Dar Gibson once informed me by e-mail during the "winter" (it was summer in the Northern Hemisphere) of 1999. "It's the largest crew ever, the usual number is around mid- to upper 20s. We have a construction crew wintering for the first time, since we are in the early stages of the SPSM [South Pole Station Modernization] plan. So the crews for the next several years will be about this large. Summer population tops out at just over two hundred, though I hear this year they are going to step it up to maybe two-twenty or two-thirty. Yikes! Three shifts of construction and lots of science personnel." Other nations also maintain scientific outposts on the harsh glacial landscape—or icescape, rather—such as Russia's at Vostok Station. The continent as a whole is larger than the United States and Mexico combined, and yet the total population never exceeds a few thousand—barely enough fill a small town.

I VISIT MARK Ross-Parent again on a sunny afternoon in New Hampshire. Dressed in jeans and an old shirt, he works in his garden till the sun slants toward the west and the shadows lengthen. The sky above us pours with blue, interrupted by billowing white cumulus clouds. A lower, darker cloud, cumulus congestus, looms over a distant mountain and sprays virga—showers of rain that evaporate before they reach the ground. Sunlight sparkles in the falling water. A few high cirrus clouds hover in the east like filaments of a spider's web.

Cirrus clouds form near the top of the troposphere, the normal boundary line of weather we can see. But the aurora looms higher, and so too does a special type of cloud that is visible from the pole and other high latitudes.

Polar stratospheric clouds, or nacreous clouds, appear in the stratosphere 12 to 20 miles above the surface of Earth. These clouds are colored like the spectrum. "They form way above the typical atmosphere," Mark instructs. "They're composed of gas, not necessarily water like these clouds," he says, pointing for comparison to the low, puffy cumulus dragging their shadows through the valleys of New Hampshire.

"The reason you can see them in polar regions is it's dark or twilightish, but the sun can still hit them because they're so high." In the northern hemisphere, identical clouds appear over Scotland and Scandinavia; you do not need to travel all the way to the poles to spot them. "They're also called mother-of-pearl clouds." PSCs consist mostly of water and nitric acid, as well as sulfuric acid.

Noctilucent clouds form at even higher levels, and are less understood. They appear at altitudes of nearly 50 miles, at the edge of space. "They're practically out of the atmosphere," Mark emphasizes. Noctilucent clouds ripple with blue-silver colors, moving rapidly across the sky on the high-altitude winds. To see them is to know you are far from the warmth of the equator. Mark tells me that they were not uncommon at the South Pole.

What *is* uncommon at the pole, he says, is something most people wouldn't expect—storms. Mark has put ice in his beard both in Antarctica and on Mount Washington over several seasons, so he should know. "Except for the extreme cold at the South Pole, overall weather on Mount Washington is much more extreme," he remarks.

"South Pole weather was pretty predictable," confirms Gloria Hutchings, a slim, dark-haired woman who spent seven seasons in the Antarctic in the 1990s. "Most of the time, I'd say probably eighty-five percent of the time in the summer, it would be sunny. And there'd only be a few days when you'd get some intense winds or whiteout conditions. But there were some great sundogs and some iridescent clouds."

Gloria Hutchings worked and lived at both the South Pole and coastal Palmer Station as a materials coordinator before arriving— sure enough—at Mount Washington. She now works at the base of the mountain for the Appalachian Mountain Club. Her last season at the bottom of the world came in 1997. "I ended up getting pregnant, so I couldn't go down to the ice anymore," she says. But she hasn't forgotten the ice. Along with her husband, Thomas, she currently maintains an e-commerce Web site, *The Antarctic Connection*, providing information, books, maps, and other products that feature Antarctica. "It's just one of those things," she says. "You get Antarctica in your blood and you just want that connection. The Antarctic connection."

WHATEVER THE SOUTH Pole lacks in stormy weather, it more than makes up for in sheer desolation—and isolation. After a long winter night at the South Pole, cabin fever sets in. Most of the crew are eager to return home.

The route from comfortable, temperate climates to and from Antarctica is much the same today as it was when Scott and other explorers first set sail; you must leave from Argentina or—more commonly—New Zealand, and return by the same route. Only the mode of transport has changed. Transportation back to New Zealand from McMurdo Station usually involves a bumpy, six-hour flight through the upper troposphere in a noisy LC-130 Hercules ski-equipped cargo jet, colloquially referred to as a C-130 or "Herc." New Zealand is the first stepping-stone on the journey home.

"I was one of the lucky few who took a ship back," Mark Ross-Parent remembers years later. "Most people just want to bolt out of there." He explains how he turned down the quicker option of air travel and instead spent seven rolling, lurching days at sea, suffering from motion sickness. When he finally arrived in Littleton, New Zealand, just south of Christchurch, it was like another world.

"Certainly the climate was a shock," he says. "I went from sixty below zero to sixty above. The biggest change that was noticeable was I wasn't wearing twenty-five pounds of clothes anymore. I felt very light."

Approximately 90 percent of people going to and from American outposts on Antarctica will pass through New Zealand. Travelers disembark at Christchurch, their faces flushed by a warm, welcoming breeze. The ice age they recently witnessed at the bottom of the world now fades away. Call it "climate shock." Ever since scientific stations were established in Antarctica in the early twentieth century, thousands of men and women have been surprised in New Zealand by the sudden embrace of moist, warm air. Off the ice, the pace of the air quickens. Suddenly, the world is once again turning beneath their feet. Day and night flash and whirl.

I know none of this from personal experience, of course, only from talking to those who do. But if the feeling of transition upon arrival in temperate lands could somehow be compressed into a few hundred words, the scene sketched below might encompass some of the highlights.

Try to imagine the moods and sensations of setting foot again in a half-forgotten temperate climate after a year of Ice-Age conditions. You are blinking and squinting in the bright sunlight of Christchurch, New Zealand, at 43 degrees 31 minutes south latitude. You lift a hand to your forehead like an improvised cap and peer at the landscape, shielding your eyes.

You have never seen the sun fly so high, not for at least a year. The daylight sizzles. Your shadow shrinks into a dwarfed caricature at your feet, dark and deep. "Strange," you think. Back in your erstwhile home of Antarctica, shadows were never so pronounced. There, the ephemeral shadows of human beings stretched thinly across the ice toward the far horizon, pale reedy giants. At the bottom of the world, the sun's rays always hit Earth at a slant—glancing blows of light like the slivers of stone a bored child skips repetitiously across the surface of a pond.

Just out of sight on the Banks Peninsula, saltwater waves crumble against the east shore of the South Island, upon which Christchurch is the largest city, with a population of 300,000. The sea is alive; whales slap and smack their fins against the surface of the Pacific, amusing the tourists. Inland, a soft breeze filters through the city's parks and green gardens. Tall mountains capped with snow rise toward the sun in the north.

The flat, abrupt daylight of New Zealand hits the ground with surprising intensity. Christchurch stands midway up the planet's southern hemisphere on a chain of islands, separated from the icy South Pole by 3,000 horizontal miles of ocean, ice, and glacier, plus another 9,300 vertical feet of rapidly thinning air. The globe spins faster here, and even quicker in the band of tropics to the north. (At the equator, the ground hurtles eastward at 1,038 mph.) As hours pass, the sky reddens at dusk and then swiftly turns black. Starlight cuts through the canopy of the atmosphere. The air is remarkably viscous at sea level—a thick fluid you can almost drink like a juice.

Michael Courtemanche, formerly of the Mount Washington Observatory, vividly remembers landing in New Zealand at midnight after 13 months at the pole. "All of a sudden you could smell all these different smells. It was kind of weird. A bit eye-opening."

His opinion is confirmed by Gloria Hutchings. When I catch up with her to talk about Antarctica, it is a warm, sunny day, very different from the weather at the pole. "At the South Pole you're just sensory deprived—which you probably know from being on the summit," she tells me. I glance up at the peak behind us and, sure enough, it is wrapped in its typical ornament of fog. The rest of the sky is blue. "It's just sensory deprivation for four months. I mean, you don't smell— well, the only smells that you get are the food in the galley, or JP8 [a fuel additive] from the airplanes."

I ask about the first impressions of setting foot back in the temperate climate of New Zealand, and she smiles. "I remember that first year, arriving in New Zealand, and it was a rainy nighttime, and we hadn't seen night in four months either. So, it was just wonderful— the smell of the earth and the grass. Someone had just gone by and mowed the lawn. We just walked for hours after arriving in New Zealand. It was great."

When Dar Gibson's plane finally landed in New Zealand at 1:15 A.M., he, too, marveled at the change. He remembers: "It was good to see the darkness again and breathe the warm, moist air. What an incredible feeling. I rode all the way to town with my head hanging out the window like a dog, nose twitching as it picked up every scent, smile across my face to finally have some wind through my short hair and across my face that didn't threaten my skin with frostbite."

BACK AT THE geographic South Pole, the air is thinner than you would suppose. The icecap juts up high into the troposphere, but at an elevation of 9,300 feet above sea level, the atmosphere is as rarefied as air at 13,000 feet would be back in the temperate climate of New Zealand. This is due to the relative thinness of the atmosphere at the poles. The centrifugal force caused by Earth's rapid spin makes the atmosphere bulge out at the equator and compress at the poles.

This compounds the effects of altitude for any new arrival used to breathing the rich air at sea level.

Just as the planet is not perfectly spherical, the layers of the atmosphere also vary in depth. At the South Pole, the roof of the tropopause lies only five miles above sea level, compared to seven miles in temperate latitudes and nearly ten miles at the equator.

Jack Halpin made only one brief stop at the South Pole during his stint in Antarctica in the late 1970s, but the rarefied air—exaggerated by the thinness of the atmosphere above the pole—hit him right away. "We were totally out of breath and had to go back to the plane to get some oxygen before we did anything else," he remembers.

Once Jack and his colleagues were refreshed by bottled oxygen and had disembarked from the plane, a South Pole tradition compelled them to jog around the world—literally all the way. "The first thing we wanted to do was go to the geodesic pole," he comments. "They have a thing down there about running around the world. They go around the pole. It looks like a barber's pole stuck in the snow," he says with a laugh.

All that's required to run around the world is a passage through a full 360 degrees of longitude and the equivalent of all time zones. At the Amundsen-Scott Station, this feat takes an average running time of two seconds. The total distance: a few feet.

Up at the equator, by comparison, the same endeavor requires a journey of 24,900 miles. In the sixteenth century, Magellan took more than three years and failed to complete the trek alive.

The South Pole station is perched atop a mile and a half of ice; the continent itself is deeply buried. In fact, the geographic South Pole changes position at a rate of one inch per day as this massive bulk of ice slides off the high plateau toward the sea. Antarctica is a land of glaciers, and a glacier is nothing more than a river made of ice, always in motion.

Curiously, at the South Pole the wind always blows from the north. But where else could it come from? No other direction exists except straight up and down. For the sake of complete records, meteorologists at the Amundsen-Scott Station divide the horizon into quadrants: a fictional east, west, north, and south. The wind's direction is determined by the line of longitude from which it came—the Prime Meridian represents grid "north." To simplify matters further, they also use a single time zone: Greenwich Mean Time.

Although winds are typically weak at the South Pole, the air stiffens with cold and the sky is dark and moody. The endless night clutches the stars in black fingers and does not let go. "You can go outside and walk around," veteran weather observer Mike Courtemanche told me after a long winter at the pole. "But you're not going to go anywhere. You could walk nine hundred miles and you wouldn't hit anything but white."

Thin air and a lack of light pollution make the stars very crisp. "There's absolutely no light pollution," stresses Mark Ross-Parent. For six months in winter, only the moon and the aurora—plus the occasional bonfire—glints off the snow and ice. But finally, in September, the sun's rays spike the sky with hints of color. "Essentially we had five months a year of pitch-dark," says Mark. "The aurora would cast some light, but it was pretty dark. We had a week or two of twilight on either side of winter."

Before the first shadow is etched onto the ice, Antarctica begins a long, cool twilight before the icy dawn.

In September, at winter's end, the daystar begins a long climb over toothy ridges of volcanic mountains near the coast of the southern continent. Mount Erebus is the largest of these peaks, overlooking McMurdo Station and the Ross Ice Shelf. Daylight arrives first at the seacoast, where emperor and adélie penguins dive in the cool waters and orcas hunt for food. But ever so slowly, the light encroaches inland.

Twilight brightens at the pole in early September. "Still cold, minus-91 right now and sunrise is only six days away!" Dar Gibson wrote those words near the end of his wintertime stint at the bottom of the world. "Believe it or not, I'm pretty used to it by now—anything above zero after this is going to seem tropical." Gibson described the incredible experience of watching the sun slowly peel off the horizon for six long weeks: "We first saw light on August sixth. Well, it was hardly light, it was more a patch of midnight blue on one part of the horizon in an otherwise black sky. From then it gradually grew brighter, with every possible sunrise color making an appearance. Now, six days before it finally shows itself, the stars are gone, washed-out yellows and oranges are the only colors, and it's so bright it seems as though the sun is already up. It's only about a degree and a half below the horizon now, probably equivalent to about five minutes before sunrise in the mid-latitudes." He concludes: "It's hard to believe that I have not felt the sun on my face for six months."

The sun circles day by day just under the horizon, tantalizing, casting an eerie glow into the upper atmosphere. At last, workers at the Amundsen-Scott station awaken to their first sunrise in half a year. The stars, once seemingly permanent fixtures in the sky, fade. Slick ice sparkles and gleams.

When Mike Courtemanche spent a winter at the Pole, he used to send a monthly e-mail report listing weather statistics and other trivia back to his colleagues in the United States. One report especially amused me. Under a column labeled "Sunrise," I instinctively expected to see a specific time of day like "4:05 A.M." But, of course, the report listed only this: "Sunrise, September 17."

Nine years before Robert Falcon Scott's untimely death on the ice, at the end of the *Discovery* crew's first winter encampment on Antarctica, he wrote in his journal, "The sunlight spreads with gorgeous effect after its long absence, a soft pink envelops the western ranges, a brilliant red gold covers the northern sky. To the north east crystals of snow sparkle with reflected light."

At the time, Scott was standing near the coast and had as yet explored little of Antarctica. The race to the pole was still a decade in the future, and most of the continent was empty and unknown. The first permanent year-round base of the South Pole—the Amundsen-Scott U.S. Research Station—was not constructed until 1957. The scientists who live there today see a landscape—and a sunrise—quite different from what exists at the coast.

In September, as the first rays of color reach around the globe and grope for the polar horizon, no peaks are visible. "There are no obvious features," says Mark Ross-Parent, speaking about the South Pole. "It's flat as a pancake as far as the eye can see."

Sometimes, you cannot see even that much. For six months at the nadir of the world, the sun never sets. Subtropical Florida calls itself "The Sunshine State," but in truth, more hours of sunlight strike Antarctica each year than anywhere else in the world.

So high is the albedo, or reflectivity, of snow that protective goggles are necessary to sightsee at the pole. The solar rays bring little warmth, but they bounce off the slick white terrain hour by hour until unprotected eyes go blind.

The first explorers of the south-polar desert squinted into the white glare from the backs of their sleds. As Scott raced to the pole on his ill-fated final expedition in 1911–12, the glare of sun on snow blinded his assistant Bowers: "I am afraid I am going to pay dearly for not wearing goggles yesterday when piloting the ponies," he wrote. "My right eye has gone bung, and my left one is pretty dicky." Although the icy climate is exactly the same today, the colloquial language has obviously changed a bit in the intervening years.

When you are snowblind, the reflected sun burns like white fire; a knife of light springs off the cold ice and stings your retinas, hot as a flame. Snow's albedo is 80 percent; like a mirror, the ice bounces sunlight back into space—and into the eyes of unwary travelers. Ultraviolet light is the culprit; it burns the mucous membranes lining the insides of the eyes. Symptoms like severe pain and swelling show up hours after exposure. Eyes fill with tears; it is impossible to see. Eyelids feel scratchy. Even cloudy weather provides no refuge, because ultraviolet light still cuts through to the surface and bounces off the snow into unprotected eyes.

Darkened goggles, apparently, are just as important as insulated parkas and warm boots. Without the latter, you freeze. Forget the former and you burn—in the eyes.

The high elevation and thin air of the interior compounds the problem. There is less atmosphere for the sun's rays to cross. The explorer Will Steger, who with five companions crossed Antarctica by dogsled in 1989, remarked on the problem of sunburn, which no amount of suntan lotion could assuage. In his essay in the November 1990 issue of *National Geographic,* he commented on how the sun "etched" reddish burns across the cheeks of one of his fellow travelers.

IT IS IRONIC that the sunniest land on the planet is an ice-covered desert that only has one "day" per year. But the icecap repels the sun's slanted rays, reflecting the heat rather than absorbing it like the dark mud of the tropics, which has a low albedo, would do. The glaring brightness that blinds the unprotected eye is also what helps keep Antarctica cold.

The South Pole is colder than the North Pole, which lies at sea level and is in fact a frozen ocean without any land. Water acts as insulation, even in its frozen state.

Antarctica contains 7,200,000 cubic miles (29 million cubic kilometers) of fresh water, locked up as ice. Theorists have suggested that the unspoiled, fiercely cold, unique environment of Antarctica would be one of the first places to exhibit the effects—if any—of global warming.

"It would certainly be a place where it would be noticed, because it's mostly unaffected by small changes," comments Mark Ross-Parent. But he states that a study of temperatures around the world is the most direct method of assessing the problem.

What if Earth really is heating up? Some people fear exactly that. Icebergs are breaking off the Antarctic coast and floating away into warmer waters. That's nothing new, but for the first time in human history, the Ross Ice Shelf and other large ice masses are showing signs of instability. Melt all the ice in Antarctica, and sea level will rise 210 feet (65 meters) throughout the world. Theorists suggest that certain changes in our atmosphere—an increase in the proportion of greenhouse gasses, for instance—may lead us on the first few steps down that road.

20

Antarctica

IN 1775, CAPTAIN JAMES COOK OF ENGLAND DISEMBARKED from his ship, the *Resolution,* and set foot on the island of South Georgia, miles below the Falkland Islands but still far from the coast of Antarctica. At the time it was one of the southernmost points of land ever gazed upon by the eyes of a European—and one of the most desolate.

"Cape Disappointment," Cook named the southern tip of the island at 54 degrees 30 minutes south latitude, once he discovered it was not the fabled Antarctic landmass which he sought.

Then as now, the island was treeless. Captain Cook remarked, "The wild rocks raised their lofty summits until they were lost in the clouds. And the valleys lay covered with everlasting snow. Not a tree was to be seen, nor a shrub big enough even to make a tooth pick." He claimed it for King George anyway.

As I read those words more than two centuries later, I marvel at how closely they echo Mallory's descriptions of the Himalayas—and Thoreau's moody condemnation of a mountaintop in Maine as alien and "such as man never inhabits." From Everest to Antarctica, some vital ingredient is the same. Tall mountains slope up into the troposphere, glaciers calve into the sea at the bottom of the world, and in both places the cold conspires to thwart life.

Along the coast of Antarctica, icy air freezes seawater into solid "pancakes" of floating ice; gelling and expanding, the ocean petrifies.

The rim of the continent spends winter encased in an impenetrable, watery cement.

Above the ice, windblown seeds and spores drift haphazardly from warmer climates to the South Polar shoreline, but they bear no fruit. Biting wind nips their buds before they can grow. Inland, the wind funnels snow in massive drifts. Cold air thickens the blood of any living creature foolish enough to wander away from the marginally hospitable climate that clings to the shore. The interior is a dead world. No land-based vertebrates of any kind can endure the weather of Antarctica. A few colonies of bacteria survive in glacial ponds and deep under the ice. The largest animals to live on land on a permanent basis are a tiny, two-winged fly and a quarter-inch-long species of midge—nothing larger. All other living creatures—including human beings—are merely visitors.

"I've seen a gull at the South Pole, about eight hundred miles from the ocean," remembers Mark Ross-Parent. "He probably died. How he made it there I'll never know."

His colleague Dave Thurlow also noted in his journal during a South Pole stint in 1988 just how rare life is in that harsh climate—and how peculiar it looks when it arrives:

> We now have wildlife at the Pole! A South Polar skua (a large scavenger bird) showed up today at our dump where the pickings are pretty good. This is the only form of life here, besides Homo sapiens, and it's kind of a kick. I can't believe it ever found this place. There is absolutely nothing but ice for at least 450 miles in any direction, and it's 20 below and there has got to be an easier way to get by.

Antarctica's temperatures may heat up to a balmy 35 degrees Fahrenheit at the coast on a summer day in January—or else plummet to the lowest temperature ever recorded on Earth: minus-128.6 degrees, measured at Vostok Station in East Antarctica in 1983. At coastal McMurdo Station, average summertime highs top out at close to 30 degrees, and the all-time record high is 49. The pole offers a sharp contrast, with average temperatures during the warmest month of summer at around minus-15 degrees Fahrenheit and a record high temperature of only 7.5 degrees.

The Russian station Vostok lies nearly 800 miles (due north, of course) from the geographic South Pole. Situated far inland atop a high plateau of glacial ice, it is the human habitation nearest to the so-called "Pole of Relative Inaccessibility." That's the location that is, on average, the farthest point in all directions from the polar sea. It looms over 11,000 feet high. At the Pole of Relative Inaccessibility, the atmosphere is perilously thin, and the sun casts little warmth but long shadows. Some call it the harshest place on the planet.

"We once had to rescue the Russians," Jack Halpin remembers, referring to a plane crash deep in the area of inaccessibility during the late 1970s. "They waited a week before they called us and asked for help, because it was still the Cold War." Halpin speculates that nationalist pride and the political climate of the day prevented the Russian scientists from seeking the aid of their Cold War enemies. Five crew members died, says Halpin. "And the rest were in pretty bad shape." To lose your way in this frozen desert is to perish. The interior resists the encroachment of life with a bone-numbing cold. The unwary or unlucky often die.

Surface temperatures on the East Antarctica plateau plummet to depths too cold for ordinary mercury thermometers to measure. The freezing point of mercury is only minus-39 degrees Fahrenheit; beyond that point, the liquid in the bulb of the thermometer hardens to ice. Therefore, if you wish to learn the temperature here to the nearest degree, and the rattling and trembling of your own bones is not an accurate enough gauge, you must use what is called a "spirit in glass," an alcohol thermometer. Spirit-in-glass can withstand temperatures down to minus-130 degrees Fahrenheit before freezing. Alcohol is considered slightly less accurate than mercury, but there is no practical alternative.

On a typical winter night at the South Pole (there are no winter days) temperatures often drop to 80 degrees below zero or colder. But no place on Earth is as painfully chilly as the area of inaccessibility around Vostok Station, site of the frostiest temperatures on the planet. By comparison, back home in the United States, Alaska holds the national record low with a comparatively mild minus-80 degrees Fahrenheit. Even Siberia bottoms out with an all-time low of only minus-90 degrees. Antarctica wins the planet's deep freeze battle by a landslide—or, perhaps I should say, an avalanche.

"ISOLATION" IS THE first word that comes to mind whenever I think about Antarctica. Trekkers to the southernmost continent encounter a strict loneliness enforced by thousands of miles of wind, ice, and cold.

When humans intrude at the pole, a hostile wind pushes snow to cover up their traces. The living are not welcome here. Artificial habitations like the geodesic dome of the Amundsen-Scott Station protect a population of a few dozen scientists and support personnel from the elements. Other than unseen bacteria, these researchers represent the only living species present at the pole on a regular basis. No birds chirp and murmur in the week-long twilight—except for the occasional garbage-picking skua, as noted by Dave Thurlow. No flies, ants, or spiders exist, except for the infrequent specimens imported with cargo from New Zealand. ("People gather around the fly and watch it, because it's so rare to see one," according to a McMurdo Station Web page.) No squirrels, foxes, or even polar bears scurry or roam across the snow dunes and rippling sastrugi.

So why do people come here? The weather allows little of what most of us desire. The heating bill is exorbitant. Wind and snow hinder colonization and render traditional agriculture impossible. The continent provides no rich harvests, no timber or other items of trade. In fact, the absence of commercially valuable items (other than the furs of seals and blubber of whales along the coast) is what kept civilization from populating Antarctica as it did all other lands. It is now believed that deposits of coal, oil, and metals do exist deep under the ice, but the technology of earlier centuries was too primitive to harvest—or even detect—these treasures. In modern times, the ice is protected from exploitation by treaty—unlike lands with similar climate and topography, such as Alaska.

The emptiness, ironically, is what motivates scientists to come here in the first place. In a refrigerated climate in which human beings and other forms of life are all but absent, a pristine environment has survived—a perfect laboratory.

In 1976, the Office of Polar Programs of the National Science Foundation provided a grant to collect and study meteorites on Antarctica. Meteorites, simply put, are rocks from outer space—usually silicate or iron—that survive the friction-induced heat of passage through Earth's atmosphere and collide with the ground. (Meteors, by contrast, are the streaks of light you see in the night sky—"shooting stars." They consist of sand- and pea-sized chunks of rock that burn up entirely in the atmosphere.)

One of these meteorites, collected in Antarctica in 1984 and designated ALH 84001, would later stir up speculation about ancient microbial life on Mars. The potato-shaped meteorite fragment weighed 4.75 pounds at the time of its discovery, and upon closer examination in 1994, it was revealed to have a Martian origin. The meteorite contained traces of gas identical to the atmosphere of Mars, a fact known thanks to the *Viking* missions in 1976, which had analyzed the thin, dry vapor that passes for air on the Red Planet. The composition of Mars' atmosphere is radically different from the atmospheres of, say, Earth or Venus, and thus provides a unique signature for the planet.

Another surprise awaited. In 1996, Dr. David McKay and his colleagues announced that ALH 84001 appeared to contain possible fossilized bacteria (and traces of organic chemical compounds from the decay products of bacteria) from more than three billion years ago—from Mars! Researchers believe that ALH 84001 originally formed from lava on the Red Planet approximately 4.5 billion years ago, and was later heated and deformed—metamorphosized. Around 3.6 billion years ago, flowing liquid water deposited carbonate mineral globules—and possibly life—in the rock. The presence of complex molecules in ALH 84001 called polycyclic aromatic hydrocarbons—PAHs—provided one clue suggesting the decay of organic matter.

The microscopic fossils from Mars—if that is indeed what they are*—are fantastically small. The alleged bacteria in ALH 84001 measure only 20 to 100 nanometers across. (A nanometer is a billionth of a meter.) For comparison, imagine the microscopic world impossibly enlarged, so that a meter (a little more than a yard) now covers the entire distance from New York to Los Angeles. On that scale, a nanometer would still be thinner than a pencil—and a typical Martian nanobacteria no bigger than a soda can. Life doesn't come any smaller.

At least 16 million years ago, possibly longer, the stony crag we now know as ALH 84001 was ejected into space when an asteroid or comet struck the Martian surface. It then drifted through the solar system for unknown millennia, until at last the tug of Earth's gravity pulled it down—to Antarctica, of all places. The meteorite struck the southernmost continent 13,000 years ago. It had traveled from the desert of Mars to the driest desert on Earth. It buried itself in the ice and waited to be discovered.

Working at the bottom of the world as a Navy meteorologist in the 1970s, Jack Halpin did not take part in, but was aware of, the project known as ANSMET (Antarctic Search for Meteorites). "The ice is all moving," explains Halpin, sitting in the lounge of the Mount Washington Observatory. Gesturing at a large topographical map spread across the table, he describes how the motion of ice in Antarctica acts like a conveyor belt, sliding the meteorites from high elevations in East Antarctica down toward the coast. He calls it "a river of ice." That's all a glacier is, really, a frozen stream flowing ever so slowly to the sea.

"It's basically a sterile continent," says Halpin, though recent discoveries of bacterial life thriving in even the harshest of environments suggest it may not be as sterile as was once thought. This frustrates researchers by increasing the potential of contamination. "One of their

---

*There is considerable skepticism on this point. New evidence and analysis suggests that the Martian "fossils" in ALH 84001 may not be true fossils after all.

projects was to look for amino acids—building blocks of life from outer space," he explains. Two decades later, the evidence of life from Mars is intriguing, but not yet irrefutable.

Only a handful of the more than 8,000 meteorite fragments discovered by ANSMET have anything to do with Mars. A majority of these fragments come from the moon or asteroids and contain clues to different astronomical riddles—including the origin of the solar system 4.6 billion years ago. ANSMET is, in a sense, the search for the building blocks of the solar system in which we live.

A cold, incredibly dry climate and a mobile ice sheet combine to make this research possible. They provide a relatively uncontaminated environment, and it also helps that black rocks from space stand out sharply against the white terrain. If a Martian meteorite brimming with evidence of alien bacteria had fallen in, say, a Brazilian rainforest, the native terrestrial life would quickly have swarmed over it and covered the traces. But Antarctica provides a deep freeze, and the moving ice continuously brings these meteorites to the white surface, where they are easier to locate.

The ANSMET Web page sums up the attraction of Antarctica to researchers:

As the East Antarctic icesheet flows toward the margins of the continent, its progress is occasionally blocked by mountains or obstructions below the surface of the ice. In these areas, old deep ice is pushed to the surface and can become stagnant, with very little outflow and consistent, slow inflow. When such places are exposed to strong katabatic winds, massive deflation results, removing large volumes of ice and preventing accumulation of snow while leaving a lag deposit of meteorites on the surface. These areas exhibit a variable balance between infall, iceflow and deflation, all of which are intimately tied to environmental change during recent Antarctic history. Over significant stretches of time (tens of thousands of years) phenomenal concentrations of meteorites can develop, as high as 1 per square meter in some locations.

JACK HALPIN'S WORK once brought him face-to-face with a perfect example of the power of preservation in the bitterly cold, dry climate. One of his duties, he explains, "was to do ice reconnaissance for ships." Two ships a year dock at McMurdo Station, bringing supplies for various outposts around the continent. "The pier's made of ice, a big block of ice. They have to fix it up each year. And it's right next to Scott's 1903 hut, Hut Point." At Hut Point, Halpin made a discovery. Leftover canned food from Scott's original expedition is still stacked inside on the shelves—and it is still edible. (I don't ask how Jack knows this, but he seems confident.) "It's as good as the day it was brought down in 1903," he insists.

He's probably right. "Outside the hut, there's also a dead seal and a sled dog, basically freeze-dried," Jack continues. The cold, dry air inhibits decay. On a typical "humid" day in Antarctica, if you took a parcel of the polar air and warmed it to room temperature, the relative humidity would reach only about four percent, and as Jack puts it, "That's real dry." It rarely if ever gets more humid.

Except for the scientists who take advantage of this natural laboratory freezer at the bottom of the world, the land is empty. In the days before e-mail, contact with the outside world was even more treasured than it is today. Nothing was more eagerly anticipated by the exiles of Antarctica than a mail drop. A former colleague of mine who once spent some time "on the ice" told me a funny story to illustrate his point. It begins with a hotshot pilot flying stunts back and forth over McMurdo Station. Then the pilot noticed the debris of an old plane crash; the twisted metal glinted in the sloping sunlight. "What's that wreckage at the end of the runway?" he queried the tower. "That's what's left of the last pilot who flew over and didn't drop any mail," came the acerbic reply.

I can only imagine what such solitude is like. Back at Mount Washington, the isolation we experience is nothing but a pale echo. Although we do live wrapped in clouds and cut off from home, civilization is always just a short (if sometimes dangerous) walk down the hill.

Still, Mount Washington is more isolated than most places. The weather is the cause. Except for studying the harsh climate, or admiring the view, there is little reason to go there. Even the view is hard to enjoy when a hurricane-force wind at minus-35 degrees is painfully freezing the water in your skin.

Once during a late summer snowfall, the summit crew and I watched a frowning TV meteorologist from Portland, Maine, point to his weather map and with a single sentence emphasize our loneliness. "Yes, it's snowing on Mount Washington," he said with a raised eyebrow. Then he shrugged. "But not many people live there."

"How true," I thought. We live here. But our detachment from civilization is only temporary; we can travel home to the valley as soon as the wind calms enough to permit a safe trek outdoors. A stint at the South Pole means true exile; in winter it is too cold for the LC-130 cargo planes to operate safely—the hydraulic fluid begins to gel. Evacuation is altogether impossible, and so a doctor must stay on hand year round for emergency appendectomies and other crises. There is no alternative. No one can ever just walk away.

"Let me just say that four and a half years at the observatory did not prepare me for the isolation of the South Pole," exclaims my former colleague Mike Courtemanche. I ask him what was hardest about his experience, and he replies without hesitation: "Seeing that last plane leave."

Courtemanche worked in Antarctica for 13 months, from 1996 to 97, and remembers how truly cut off from the world the Amundsen-Scott station was during the dark winter. "More than likely if they had to land a plane, it probably wouldn't take off again. Below minus-sixty degrees the hydraulics freeze, and I've heard that the skis freeze to the ground. They say in an emergency they'll come and try to evacuate, but you know that probably wouldn't happen."

Two years later, in June 1999 at the height of the Antarctic winter, that indeed proved to be the case. A woman working at the Pole—later identified as the station's only doctor—detected a possibly cancerous lump in her own breast and required emergency medical attention. The "hospital" and equipment at the South Pole was not intended for such a crisis, but evacuation was out of the question. The weather conditions simply didn't allow it. Instead, a risky airdrop of medical supplies was attempted.

By coincidence, another former Mount Washington employee, Dar Gibson, was working at the South Pole as a meteorologist at the time. He participated in the retrieval of the airdropped supplies, and later wrote to me via e-mail:

"We heard the airdrop created somewhat of a news stir back home. It was a pretty exciting event, something we definitely did not expect this winter. A lot of what we heard from home and read on the news Web sites was relatively inaccurate, but I guess that's to be expected. One story said we only had five minutes to retrieve the boxes or we'd freeze to death. Nope. We were out there about four or five hours total, took about two hours to retrieve all six boxes and get them inside. Got a big box of fresh fruits and veggies, that was a treat."

SOME PEOPLE WHO visit Antarctica never go home at all. Captain Robert Falcon Scott was neither the first nor last casualty of Antarctica's weather.

As high pressure sinks over the continent and spreads across the ice, downslope winds push a blinding snow toward the coast. A "Herbie" results. That's what killed Captain Scott on his ill-fated return from the pole in 1912. He perished, hopelessly lost and unable to move, only 11 miles from a food cache that might have saved his life. Trapped for days on end by whiteout conditions, Scott and his crew could not see well enough to travel. No up or down or sideways exists in a whiteout—only snow. So Scott and his companions huddled in their tents, cold and starving, until they died.

One of Scott's crew, already doomed by frostbitten feet turned to gangrene, stumbled out of the tent one night. He told the others he did not expect to be back soon. "I may be some time." They never saw him again.

Scott's only hope was to wait for the wind and snow to subside, but it never did. The swift plunge of air from high to low was like the thrust of an atmospheric knife, cutting them off from their only life-line—the hidden food cache. The storm did not permit any escape. Scott finally wrote in his journal: "Every day we have been ready to start for our depot 11 miles away, but outside the door of the tent it re-mains a whirling drift."

A few lines later, he concluded, "We shall stick it out to the end, but we are getting weaker, of course, and the end cannot be far. It seems a pity but I do not think I can write anymore."

A search party found Scott's body in the tent eight months later. The captain had pulled open the flaps of his sleeping bag, inviting in cold air to hasten the end. The tragic fate and last days of his expedi-tion is vividly described by one of Scott's original crew members, Aps-ley Cherry-Garrard, in his 1922 book *The Worst Journey in the World*.

In whiteout conditions, eyes are useless, and the howling wind deafens ears so that any cries for help go unanswered. Trapped in a blizzard at McMurdo, a man who was too bashful to relieve himself inside the one-room shack where he and three associates were con-fined risked a trip outside. He found privacy only a step or two from the door—too much privacy. Blinded by the blowing snow, knocked about by gusts of wind, he lost his sense of direction. Stumbling away from the door, he groped frantically for shelter. His hands touched air. "Too far," he thought when the shack did not reappear, so he turned on his heel and walked in the opposite direction. When that failed, he desperately paced in a widening circle. But he never found the door again. He froze to death in the blizzard, just a few steps from safety.

That particular tale may be apocryphal, but it is believable. Glo-ria Hutchings, who once spent seven seasons in the Antarctic, de-scribes the ordeal of her husband, Thomas, when a powerful Herbie rumbled through McMurdo in the 1990s: "It was total whiteout con-ditions for three days. I missed that one. Everyone has some great sto-ries about it—the ropes, going across building to building. People couldn't . . ." She stops abruptly. "Well, they weren't *supposed* to go to work."

Her own close encounter with weather occurred not at Mc-Murdo or the pole, but at tiny Palmer station along the coast. "I wanted to go for a hike up the glacier, and it was whiteout conditions, and I was like, 'Okay, I'll just wait a few hours or whatever.'" Half an hour later, Hutchings remembers, the sky started to clear. "So I got all my gear together, got ready, and then it turned back into whiteout. It was kind of an off-and-on thing." Later, the sky cleared again, and looked like it would stay that way. "I finally went outside, it cleared up enough so I went for my hike, and I got stuck in the whiteout. And I've never been in one that was just like that." She becomes animated at this point. "It just swept in and would hover there for a while. So I just hunkered down and didn't know which way to go. So I just waited it out. Winds were just howling. I guess I wasn't expecting it."

Halfway around the globe in the winter of the northern hemisphere, blowing snow has blinded me and many others in exactly the same way. It's startling how quickly you lose yourself. While driving a SnoCat down the Mount Washington Auto Road, a man named Phil Labbe once stepped outside the cab to brush ice off the windshield. Fog and blowing snow cut visibility to less than two feet. A gust jolted him, and suddenly the SnoCat had disappeared. Phil knew it was there, somewhere, just beyond the reach of his outstretched arms. But he did not know in which direction. So he guessed—and he guessed wrong. His hands touched empty space. He guessed again. No luck. The wind continued to buffet him. Hours later he walked to safety at the base, having met on the way the search party that had come up to find him.

I've had my own encounters with whiteouts. I once walked into a eight-foot-tall wall of snow that just six hours earlier had not existed. It was time to collect the precipitation can, only 100 yards from the front door of the observatory. But that was far enough. As I climbed over the hard-packed ridge of drifted snow the wind screamed in my ears, and I wondered if I was lost. In my case, of course, the nearest food cache was a well-stocked refrigerator a mere flight of stairs away; but for all that I could see, it might as well have been at the North Pole.

The sky disappeared. When by blind luck I stumbled back to the door, I spoke to a colleague who had followed me a short distance outside. He was waiting by the door. "There's a mountain of drifting snow out there, but it's so white you can't see it," I gasped.

"I know," he said. "I saw you stumble and fall there but didn't know why. Until I hit it too."

Sarah Curtis recalls her own frightening encounter with the blinding snow. She, too, had left the safe confines of the observatory to collect the precipitation can—a routine task we must do many times, every six hours. But this time, thick fog and blowing snow enveloped her. "You know we have our route to get out there," she tells me. "You go to the rocks and then you go to the Tip-Top steps and have like twelve or sixteen paces to the precip can. And I took about two paces away from the Tip-Top steps and turned around and they were just nowhere to be found." Suddenly, she was blind. "That blew my mind, it was such a whiteout. Part of you is saying, 'Well, no matter what direction I go in I'm going to run into something that I recognize—a building, the steps, the precip can, the Cog tracks.' But then, another part of me was saying, 'There could be a cliff over there.' And so you're really frozen, saying, 'What do I do?' Because your mind is playing tricks on you."

THE ABILITY TO see is something we all take for granted until it is suddenly taken away. Not only wind, but blowing snow, cloud banks, and fog are all hazards to airplanes attempting to land at McMurdo Station and elsewhere in Antarctica. Fortunately, crashes are rare, and it was Jack Halpin's responsibility in 1978 to make sure they remained so. "Our biggest job was to give pilots flight winds so they'd know how much gas they were using," he remembers many years later. Ski-equipped LC-130 cargo planes provide transportation to and from Antarctica; their top speed is 225 knots, which seems plenty fast to me. But Halpin remarks, "They're kind of slow, and any wind slows them down even more."

Ride a bicycle into a strong headwind and you will understand his last remark. On the back roads of New Hampshire on summer afternoons, I have often found myself pedaling furiously into a 15-knot headwind on a gentle downhill slope, and still going nowhere. Yet on the same stretch of road with a tailwind, it's almost possible to coast uphill. For aircraft, with their large surface area and tremendous speed, the effect is exaggerated. A passenger jet flying from New York to London with the jet stream at its back traverses the Atlantic more easily than a plane that must battle the river of wind head-on for the duration of the western passage.

At McMurdo Station, LC-130s and C-141 "Starlifter" cargo planes fly in periodically from Christchurch, New Zealand, landing on sea ice from October through mid-December. "The season was pretty short," Halpin recalls. "They had to stop the flights because the ice would break up." The runway stretched two to three miles across flat, frozen ocean water.

Then as now, most traffic to Antarctica involves military planes flying scientific personnel to the outposts, but tourist planes also visit the southernmost continent. Despite the alien, unearthly landscape down below, pilots must still contend with exactly the same weather hazards that exist everywhere else on the planet. Like all mountains, the volcanic peaks of Antarctica trigger fog, wind, clouds, and precipitation. Air surges up their slopes, cools, and condenses.

In November 1979, five years before the discovery of the famous Mars meteor, Jack Halpin witnessed the aftermath of a plane crash near the coast. The culprit: fog.

The victims were sightseers who ended up seeing more than they bargained for. In a normal year, Jack Halpin states, four to five flights from New Zealand and Australia buzz over McMurdo Station. And although most passengers are tourists, occasionally a VIP will tag along as well.

"Just prior to the fiftieth anniversary of Admiral Byrd's trip, we had a flight of dignitaries," Jack remembers. Guests on that day included the grandson of Admiral Byrd himself, now an admiral in his own right. The first Admiral Byrd first flew over the pole in 1929, celebrating technology's conquest of the elements, an adventure recounted in great detail in the pages of *National Geographic*.

While Admiral Byrd's grandson and other dignitaries toasted that accomplishment, another plane—a tourist flight from New Zealand—arrived overhead. "It was cloudy on the back side of the mountain chain," recalls Jack Halpin in a monotone. "Ross Island is a composite chain of volcanoes, Mount Erebus being the biggest one."

Since it is so easy to get lost in fog, certain safety measures were in place. "There was an unwritten rule," says Halpin. "They weren't supposed to go below sixteen thousand feet if there was fog or snow in the area. That would keep them above any peaks, even if they couldn't see them."

For unknown reasons, the pilot ignored those instructions. "He tried to descend below the clouds, but that didn't happen. He was in the fog all the time."

Halpin speculates that the flight crew may have tried to drop below the cloud deck, hoping for better visibility. In the meantime, the plane descended dangerously low. Instead of finding clear air beneath the clouds, the plane hurtled into a mountain and exploded in flames. No passengers or crew survived.

Halpin joined the salvage team in the hours and days that followed. "You could see wreckage in the ice. Where the plane hit, it had bounced twice before the engine exploded." The anniversary of Admiral Byrd's feat had suddenly taken a grim turn.

A writer from the *New Yorker* magazine was also present that day, intending to cover the celebration of Admiral Byrd's flight. Suddenly she found herself with a new and unwelcome topic. "She didn't know how to handle the disaster. That wasn't what she was there for," says Halpin.

No one knows for certain what caused the tragedy. According to Jack Halpin, the pilot's inexperience may have played a role. "They retrieved all the film from the tourist cameras. Mount Byrd was visible, and that should have clued him in right away that he was on the wrong side of the island." He pauses. "None of the crew except one of the flight engineers had ever flown to Antarctica before."

A total of 257 people died in the crash, Halpin remarks. Then he adds, "Make that two hundred and fifty-eight. The fiancé of one of the stewardesses committed suicide a few weeks later."

WATER IS A curious thing. It liquefies into a substance called fog
and floats in the air, blinding pilots and disguising rocky mountain
peaks in a misty veil. When cooled, it hardens itself into a rock called
ice—a rock which, almost uniquely, is denser as a liquid than as a
solid. That's why an ice cube floats in a glass. That's why icebergs bob
on the surface of the ocean instead of sinking to the bottom.

The ice and the savage climate are what kept would-be discover-
ers of Antarctica at bay (sometimes literally) for centuries. Antarctica
is a land like no other, but the hidden forces of weather obey the same
rules of meteorology and are—to an extent—predictable. Air circu-
lates constantly between the pole and warmer regions near Earth's
equator. Cold, dense, sinking air cascades from the sky over the high-
pressure region of the pole and quickly spreads north toward sunnier
climates—where it once again rises, cools, and sinks. The cycle repeats endlessly.

Warmer air percolates upward at the equator and then spreads
across the sky at high altitude, gradually cooling until it descends
again in areas of high pressure. Bands of high pressure exist perma-
nently at latitudes of approximately 30 degrees ("the doldrums"), 60
degrees, and, of course, at the two poles. But in between, countless
ribbons of wind undulate and coil at varying altitudes. All around the
globe, the wind loops and swirls. The jet streams are the most obvious
examples; these high-altitude rivers of wind help spawn storms, determining where they will go and dispersing air masses from one land
mass to another. The wind never rests.

North of the Antarctic coast, storms may spin around the globe
for a full 360 degrees of longitude, carrying parcels of air for thousands of miles and driving the West Wind Drift. These tempests frequently lash against McMurdo Sound, bringing a warmer surge of
moist wind to the coldest, driest continent.

"Storms along the coast at McMurdo were really violent," says
Mark Ross-Parent. "We had a lot of wind and poor visibility. Eight or
nine or ten days out of the winter we had winds over a hundred miles

per hour. A lot of blowing snow. Temperatures were not so cold—just the opposite of the South Pole, where temperatures are very cold but there's not much wind."

Contrary to popular belief, bathroom sinks and toilets do not spin one way in the northern hemisphere and another in the south. The quantity of water in the average sink basin is far too small to be influenced by the Coriolis effect; water gurgles down the drain any which way it pleases. But take a giant low-pressure system hurtling around the planet and sucking moisture from the uninterrupted band of ocean north of Antarctica, and now you have billions of tons of water to take into account. The laws of physics push and pull this watery mass in predictable directions. The Coriolis effect bends wind to the left in the southern hemisphere and spins the cyclone like a top. Evaporation of water vapor off the southern seas pumps the storm full of fuel. The sky howls.

"We had horrendous winds up to a hundred and thirty miles per hour with low visibility. Strongest I've ever seen at sea level," says Mark. The air is denser, at sea level, and so the wind packs more of a wallop. Greater numbers of air molecules press against each square inch. A gust of 130 mph at sea level hits you like a battering ram, stronger than winds of the same speed at a high elevation like the peak of Mount Everest—or even Mount Washington. "Being a meteorologist, just knowing the winds are at a hundred and thirty at sea level is impressive."

Over time, every one of these molecules of air will scatter from the pole to the equator and back again, siphoning an alternating current of hot and cold. Without this constant thermal exchange, Antarctica would progressively grow even colder than it is today, forever losing more heat than it gained from the sun's weak rays. The continent's snow and ice reflect the feeble, slanted light of summer off a slick white surface that's like a mirror. This produces a blinding light and the danger of burning your retinas, but that is only one effect. Meteorologically speaking, a far more critical result of the reflectivity of ice is this: it bounces sunbeams back into space.

Only about 13 percent of all incoming solar radiation gets absorbed by the sheen of snow. Most of the rest of that energy ricochets off the icecap and disperses into the cold cosmos. A relatively small percentage of heat is absorbed by the ice and the air, adding a slight warmth to the passing breeze. But for the most part, heat in Antarctica trickles in from the rest of the globe, from regions where the sun is more effectual and the overall climate warmer. That is all that keeps Antarctica from the clutches of a truly unfathomable cold, a nightmare climate almost as frigid as the deserts of Mars.

Whatever questions we ask about the weather, the answer (to paraphrase Bob Dylan) is always blowing in the wind. Wind brings to the pole the dispersed fragments of Caesar's breath and the seeds and spores of greener lands. Not even the wide ocean is a barrier to wind-carried spores. But when the wind finally stills, these seeds and pollens settle on a land too hostile for life. The cold and the ice stops them from taking root.

In stark contrast to the desolate interior, however, plant and animal life actually thrive along the outer coast of Antarctica. Colder water is denser than warm water and holds more dissolved gases—like oxygen and carbon dioxide. More nutrients exist than in tropical seas. The ocean is rich in plankton, saturated with photosynthetic creatures pumping oxygen into the air above the waves, helping to make the atmosphere's composition what it is today. Plankton is the bottom rung of the food chain, supporting the weight of many species higher up the ladder.

In 1994, Gloria Hutchings noticed the difference as soon as she switched from the South Pole to Palmer Station along the coast, where her responsibilities included coordinating the loading and unloading of supplies. "Palmer's the smallest station, and it was very different logistically, because everything arrives by boat. And the weather and climate—everything was different there. More wildlife. It's called the Banana Belt." She smiles. "And after the pole it did feel that way. It was a nice change."

"Once the ice went out, we saw all kinds of whales, seals, penguins, a lot of bird life," a veteran of McMurdo station once told me. The coast of Antarctica is alive. Adélie and Emperor penguins waddle comically on the ice; they toboggan on their bellies downslope into the water.

Out among the waves, right whales poke their snouts above the water for breath. Colorful fishes slip and dart beneath the ice. But only a thin rim of life circles the inner ice. Go just one mile inland and everything changes; the land is dead. Except for tough lichens on boulders and colonies of bacteria too small to see, the interior of Antarctica is barren. The cold is too extreme, too permanent during that six-month-long nightfall known as winter at the pole.

AT THE SOUTH Pole, Earth spins so sluggishly that you can crawl on your hands and knees and still easily outpace the rotation of the globe. Stars spin in a lazy pirouette in the sky. The world, at least according to northern Eurocentric mapmakers, is turned on its head.

None of that changes the forces of weather. Wind still flows from high to low in a reassuringly familiar pattern. An altocumulus cloud hovering over the Amundsen-Scott South Pole Station looks much the same as an altocumulus cloud in the skies of South Carolina. Storms are rare at the pole because cold air keeps the sky dry, virtually devoid of moisture.

Very little precipitation falls each year, according to Mark Ross-Parent. "We occasionally had diamond dust precipitating, but that can fall out of a clear sky." Diamond dust is also called ice fog, ice prisms, or ice crystals. It tumbles and glides ever so slowly out of seemingly empty air. In bitterly cold temperatures, tiny unbranched ice crystals condense and fall. The same phenomenon can be seen in Alaska and Minnesota, or anywhere where the conditions are right. "Occasionally," Mark adds, "we had snow grains on the warmest days, but not what you would think of as a normal snowfall."

Despite the desert-like climate, a mile and a half of snow and ice covers the pole—as much as covered the city of Boston during the last ice age in the Northern Hemisphere. The Amundsen-Scott Station is perched atop an icesheet 9,300 feet above sea level—so high that new arrivals at the pole often must inhale bottled oxygen until they become acclimatized. Headaches induced by altitude can throb painfully in the craniums of newcomers. The bulk of that solid landmass beneath their feet is ice. It had to get there somehow.

Drifting snow provides one answer to this riddle. Wind pushes snow from one end of the continent to another over countless centuries, and as a result, it steadily accumulates. The current South Pole weather station features a snowfield in which 50 stakes are placed in a rectangular pattern. "We'd measure the depth on the first day of each month. Then we'd reset the snowstakes, once at sunset and once at sunrise—that's twice a year, six months apart," explains Mark. "And that was the only way we could tell the rate of drifting."

Approximately nine inches of new snow accumulates each year at the Amundsen-Scott Station. "That's why the station has a problem being buried," Mark tells me. He notes that the first South Pole station, built in 1956–57, was completely covered by 1970. "It's still there, about twenty to twenty-five feet under the snow now. It's dangerous to enter." A second research station was constructed in 1972, "And that's being dug out now."

Photographs of the South Pole station reveal a slope of snow extending up one side, a white wave soon to engulf the entire building with a slow-motion splash. That side of the geodesic dome is a smooth white hill. The other side is silver and black, artificial and exposed. "They've excavated that one a couple of times now. It drifts on the lee side, as you would expect. The debate now is whether to build a third station or dig out the current one and enlarge it."

IN AN ERA of shrinking budgets, the fact that any debate at all exists about how to rebuild or repair an expensive scientific outpost on the most desolate spot on the planet is an indication that the questions being asked there are important ones. Science is looking for much more than meteorites in the wasteland of Antarctica.

Geology, astronomy, biology, paleobotany, and meteorology, to name just a few, are all active sciences on the southernmost continent. Meteorologists launch weather balloons daily, and dozens of other ongoing research projects simultaneously expand human knowledge and propel the careers of the scientists who oversee them. One concern which merits serious study is that a change to the climate of Antarctica may herald a greater worldwide change. The exact results of a global rise in temperature are as yet unknown.

Sudden changes in weather took the lives of Scott and his companions. Meteorologists fear that gradual changes in the weather of Antarctica—such as the melting of the Ross Ice Shelf—may indicate a greater global climate change just now beginning. How will this change weather patterns, hurricane tracks and frequency, El Niño and La Niña events, and ocean currents like the Gulf Stream? These are some of the questions being asked.

Climate is defined as the average weather pattern of a region taken over the long term, from decades to centuries. The global warming theory is that heat-absorbent gases in the atmosphere—carbon dioxide is one example—will heat Earth like a greenhouse, changing the overall global climate. Newspapers often splash the horrifying consequences across the headlines—rising oceans, islands disappearing like Atlantis, Disney World flooded, et cetera. The truth is a little more complicated.

If you could tweak the global thermostat just a few degrees, you could start a new ice age—or cook the world with global warming. 18,000 years ago in the Pleistocene, at the height of the last great ice age, the average worldwide temperature dipped by about ten degrees Fahrenheit. A study of ice cores—Antarctica provides a key location for this sort of research, as does Greenland—and fossils indicates that this worldwide cooling occurred. A change of equal magnitude in the opposite direction is what many meteorologists now fear.

"Looking at the data and looking at the trends, I think there has been global warming," says Sarah Curtis. "There *is* global warming. I haven't researched it enough to say, by this many degrees or this many years, you know. But just looking at the trends over the past hundred years, or fifty years even, in different places, it seems to me there is definitely global warming."

She pauses a moment, coming to the difficult part. "I think the argument comes from exactly why there's global warming. And I'm not going to take either side. Because I just don't know enough about it. It could be manmade, it could be natural. It could be a combination of them both, and I think that—now I'm getting opinionated—for one side to completely argue their point would be . . . naïve, maybe, at this point. Because there's so much we don't know. I think, to get an accurate portrayal, we need hundreds and hundreds of years of data. Maybe they have it, I'm not sure. Until they can come up with that, whether it be from ice-core sampling or different ways, I think for anybody to make up their mind right now about why it's happening would be premature."

The controversy over the potential causes of global warming and its political repercussions could fill a book—but it won't fill this one. Forget for a moment about politics, about problems and proposed solutions. Instead, consider how heating up Earth affects weather.

Today, the planet's average surface temperature is about 59 degrees Fahrenheit (15 degrees Celsius), the equivalent of a mild autumn day. That number is the mean, from icy Antarctica to the balmy Bahamas. But many scientists predict that the planet's average temperature will rise over the next century or so to perhaps 63 or 64 degrees Fahrenheit or even warmer.

To most people, a handful of degrees doesn't sound like much. If you walked outside on an autumn day and the temperature jumped from 59 to 63, you would scarcely notice the difference. Taken on a global scale, however, even a small bump in the thermostat has enormous impact. For one thing, any rise in global temperature would make umbrella manufacturers very happy. A warmer global temperature means more heat and energy in the troposphere, and that in turn increases the rate of evaporation off of oceans, lakes, and rivers. (Anyone who has ever hung wet clothes out to dry knows that they dry a lot faster on a hot sunny day than a cool damp one.) But as evaporation surpasses condensation, moister conditions prevail. The increase in evaporation off the oceans brings an old adage into play: "What goes up must come down." Eventually all that water vapor must condense into clouds and fall back to Earth as rain, drizzle, snow, sleet, or hail.

A few of the predicted effects of global warming actually cut against the grain of common sense—at least at first glance. Ironically, scientists now believe that a rise in Earth's temperature could actually make the icecaps at the North and South Poles grow *larger*. Or thicker. Robert Falcon Scott himself suggested this curious result back in 1903, long before the terms "global warming" or "greenhouse gas" had entered the lexicon on the evening news.

At first, it seems to defy common sense. You don't grow an ice cube by putting it in a frying pan; you expect it to melt. But as Scott realized, temperatures near Earth's polar regions would still remain very cold—often below the freezing point of water—even if Earth's overall temperature went up. Precipitation would increase globally. Though it might fall as rain in tropical and temperate climates, in the frigid interior of Antarctica it would still fall as snow. And the increase in temperature and humidity simply means it would snow much more. Only an inch or two of precipitation falls annually at the South Pole at the present time, but that amount could double or even triple, increasing the thickness of the icesheet at the heart of the continent, even as glaciers melt and calve at the coast. (Ice will persist year round anywhere that the average annual temperature stays at 14 degrees Fahrenheit [minus-10 Celsius] or lower.) Icebergs might sprawl across the seas, but the pole itself would see more snow than ever before.

THE MOST DIRE predictions of global warming offer a grim scenario. The strong El Niño phenomenon that tormented California in 1997 was a taste of what warmer, wetter weather can do—at least in a localized setting.

One of the 1997 El Niño's side effects—beyond the slackening of easterly trade winds and the increase in warm surface waters nudging against the coast of Peru—was a shift in the location of the semipermanent Pacific High to the west of California. As a result, storms moving across the Pacific Ocean suddenly changed their courses.

In a normal year, southern California is shielded to a degree by this area of high pressure. During the El Niño year, the high moved, and low-pressure systems shifted their familiar storm tracks as a result. A deluge was the next result. Storms ordinarily deflected by the high hit the California beaches full-force.

While listening to the radio one summer afternoon, I heard a lament from a desperate southern Californian man caught in a downpour. "You know you're in trouble when you've got ducks swimming in your driveway," he complained.

Heat is energy; add more heat to Earth's atmosphere and you make the sky more volatile. Both summer and winter turn violent.

A Swedish scientist named Svante Arrhenius coined the term "greenhouse gases" in 1896. At the time, he intended only to explain how carbon dioxide keeps Earth warmer than it otherwise would be. Politics and doomsday scenarios did not enter into his studies. As Arrhenius explained, "greenhouse gases" are actually a good thing. Water vapor and carbon dioxide occurring naturally in the atmosphere keep the global thermostat at a comfortable level; without them, the planet's 59-degree average temperature would plummet close to zero degrees Fahrenheit. Perhaps a narrow band of life would survive at the equator, but the rest of Earth would turn into a barren ice world, little different from the interior of Antarctica today.

The planet receives heat both from the sun and from the atmosphere, and the so-called greenhouse gases (they do not really suppress convection, which is how a true greenhouse works, so the term is misleading) are the reason why. As heat rises off Earth's surface, a small percentage of it is absorbed by gaseous molecules in the air, rather than escaping into space. Carbon dioxide and water vapor both absorb heat well; later they emit heat, and some of that energy is radiated back toward the surface. Without the heat radiated from the atmosphere, Florida at ground level would soon resemble the South Pole.

This leads to an obvious question: If greenhouse gases are good, even vital, then what's all the fuss about? The answer is that "global warming" is the fear of too much of a good thing. It's possible to take the greenhouse effect too far—and the atmosphere of our sister planet, Venus, is a picture portrait of what can go wrong. Venus is believed to be a greenhouse scenario run amuck. The boiling sky is so hot that water at the surface cannot exist as a liquid. Even the rocks at the surface soften under intense heat and pressure. Although for decades Venus was depicted in science-fiction stories as a tropical paradise, astronomers learned the truth in the 1970s, when the crushing weight of the Venusian atmosphere destroyed the first Russian probe to land on our planetary neighbor's surface. The probe's dying instruments radioed back to Earth the fact that the average temperature on that distant, cloudy world is over 867 degrees Fahrenheit.

IN ANY DISCUSSION of global warming, a rise in sea level is always ominously predicted. If the overall temperature increases to a certain level, one of the effects will be the flooding of Florida and various tropical island nations. Only a slight jump in heat is necessary before the oceans swallow these islands like modern-day Atlantises.

The cause is twofold. Glaciers in Greenland and especially Antarctica currently hold much of Earth's fresh water locked up as ice. If all of the Antarctic icecap melted at once, sea level would rise 210 feet, flooding many coastal cities and virtually all of Florida. You could navigate the streets of Miami in a boat, or perhaps a submarine. Melt just the icecap of Greenland, and a 30-foot rise results. If all glaciers in the world melted, including mountain glaciers and the Khumbu Ice Fall at the base of Mount Everest, sea level would rise a total of 250 feet. The movie *Waterworld* was no doubt inspired by just such a nightmare, though it exaggerated the results. Still, in Washington, D.C., only the tip of the Washington Monument would protrude above the waves, a forsaken crag looming above a drowned city.

However, melting glaciers are not what we should worry about. There's more to the story.

In a typical global-warming scenario, much of the rise in sea level comes not from ice, but from the thermal expansion of water. Consider: if you heat up the air in a balloon, the air expands, occupying more room until at last the balloon pops. A liquid like ocean water behaves in the same way, swelling when heated and shrinking when cooled. (So does a thermometer.) Imagine countless billions of water molecules jammed "shoulder to shoulder" in the ocean, pressed together by gravity. Apply extra heat—which is really a form of kinetic energy—and that suddenly gives them the oomph to shove their neighbors away, pushing for just a little more room. The water expands, and sea level rises. People who live along the shore either move inland or learn to snorkel.

If global warming predictions prove true, large chucks of coastline will someday soon take a dip in the sea. But what is the cause? Human or natural? Or both?

It has happened before. The climate we enjoy today is by no means the standard for planet Earth.

THAT MAY TWO decades ago during the blizzard-prone 1970s, when it snowed on my birthday, newspapers and television specials warned of an impending ice age. Nobody—at least, not in the general public—had yet heard of global warming.

The fear then was endless winter. Glaciers would roll through the streets of Boston and push Fenway Park into the Atlantic. Headlines heralded the impending march of blue glacial ice down the streets of Chicago. Meteorological soothsayers and easily excitable science writers stirred our emotions with visions of the end of civilization: crushed by a wall of ice stretching across Canada and the northern United States. I half-expected to hear about live woolly mammoths discovered in Saskatchewan.

To add weight to these fears, record snowfalls did indeed occur in the 1970s. And as for the impending ice age—well, it had happened before.

Nearly 18,000 years ago, continental glaciers more than a mile thick covered much of North America. They contained a large percentage of the world's fresh water, forcing sea level to drop 400 feet lower than it is today. The present site of New York City was more than 100 miles from the coast. The familiar outlines of continents were altered almost beyond recognition.

Currently we live in an interglacial period, with the possibility—even probability—of a return to ice-age conditions within several thousand years. "I don't buy this global warming stuff," I overhead a climatologist say recently. "I think we're due for an ice age."

Perhaps. Unless, first, a form of human-induced global warming takes us to the other extreme.

On the geologic time scale, a few thousand years is like the blink of an eye. But ice cores from Antarctica and Greenland indicate that sweeping changes in Earth's climate have sometimes occurred with fantastic speed. By studying ice cores, researchers have determined that the climate can and does change quickly and dramatically; sometimes a mere century separates a mile of ice from rich green grass and peach trees. Sometimes even fewer years are required. The Pleistocene era yielded four major glaciations, each lasting many millennia. Perhaps we are due for another.

Some climatologists have even suggested that a substantial increase in atmospheric carbon dioxide—a key "greenhouse gas"—could eventually result in a *cooler* global climate. It's all a matter of equilibrium—checks and balances. A jump in temperature would cause increased evaporation over the ocean, more water vapor in the air, and, one would think, more clouds. "Marine stratus clouds are potent cooling agents," explains Jacob Klee. "They reflect a lot of incoming solar radiation back to space." Because they are so close to the surface, however, these clouds radiate away nearly as much longwave energy as would the open ocean on a cloudless day. In colloquial terms, that means they don't "trap" heat very well. Compare two satellite images—one visible, the other infrared—and you will see the difference. In a "normal" satellite image, low clouds appear a bright, blazing white—they reflect a lot of light. But in infrared images, you can barely distinguish these clouds from the ground (or ocean) surrounding them. In this case, the clouds and Earth's surface are approximately the same temperature and emit the same amount of outgoing radiation, while the clouds' high albedo limits incoming radiation from the sun. The result: a possible cooldown. That's one hypothesis. On the other hand, a slight modification of the overall warming trend may be the only result. No one knows for certain.

The mechanics behind sudden changes in climate—past, present, or future—depend on the original source of all weather, the sun. Solar output varies slightly during the 11-year sunspot cycle; at times the sun is hotter or cooler. It is believed that deep in the Archeon Eon, billions of years ago, when the first primitive photosynthetic life struggled to pump oxygen into the primeval atmosphere, the intensity of the sun's output was only 75 percent of what it is today. Other factors include Earth's wobbling in its orbit, which effects slight changes over time. Ocean currents also migrate and reorganize the flow of air above them.

No matter how much we learn, it seems that we do not know enough. The number and complexity of forces at work in the atmosphere defies any absolute predictions. The only certain thing about the weather is the fact that it's always changing.

WIND ALWAYS TWISTS and turns, spinning storm systems the size of small nations from one horizon to another. The lungs of the planet exhale convective bursts of warm air upward at the equator; eventually the same breezes cool and fall, and howl across the poles.

Billions of tons of water vapor percolate like tiny, invisible bubbles off the world's oceans every hour, spreading clouds in all directions. Water, like wind, circles the globe endlessly. Jet streams loop and meander in the sky like indecisive rivers.

Far below the tropopause, ocean currents curl against the dry continents, bringing alternating doses of warm and cold. On Greenland, wind and water once permitted Norse settlements to thrive; two centuries later the oozing cold water of the Labrador Current changed the local climate and drove them all back to Europe, or else starved any stubborn holdouts and buried them in the snow. The Labrador Current brought a chill that lasts to this day at the beaches of Maine.

History tells us that the weather was once balmy enough to allow wine grapes to grow in Britain. Such is no longer the case. At the other extreme, a mile-high sheet of ice used to flatten an uninhabitable land that is now called Chicago.

Not too many millennia ago, Earth's average temperature dipped low enough to roll gargantuan sheets of ice down the northern hemisphere, gouging into the landscape giant holes that later filled with water—the Great Lakes. Such is the power of weather.

Today we worry about the climate swinging to the other extreme. If global warming occurs, perhaps maple trees and maple syrup will someday vanish from New England except as an exotic import. After all, wine grapes have vanished from Britain—and Viking gardens have certainly disappeared from Greenland. The maples will be crowded out by invading plant species from the south, plants better adapted to a hotter, more humid climate.

I will miss my neighborhood maple syrup, and the red tint of leaves in autumn, and the winter landscape cloaked in snow. But however fickle the climate, I take comfort in the fact that the rules that govern weather will always stay the same.

This morning, as a breeze swipes across my face in Berlin, New Hampshire, I can almost visualize the stacked columns of the atmosphere rising over my head. A filament of cirrus cloud delineates the jet stream and the top of the troposphere, above which the temperature of air steadies and even starts to rise in the stratosphere, heated by the influx of ultraviolet radiation from the sun. Much lower, a patch of cumulus clouds billows and glides on the wind.

In the condensation of a single cloud I see one link in the long chain of meteorological events that leads water to rise invisibly off the oceans, later to fall as snow on the mountains, gurgle down rivers and streams, and at last pour back into the sea.

The muggy air in New Hampshire this morning leaves me fully ready to believe in global warming. How else could it get so hot and humid? (Admittedly, the acclimatization of more than five years on the mountain may have tainted my point of view. "Anything over sixty is unbearable," I sometimes complain aloud—and people look at me strangely when I do.) I wish halfheartedly for the refreshingly cool embrace of an ice age—anything to counterbalance this sweltering air.

In my lifetime, most probably, neither an ice age nor a heat wave will occur full-force. But you never know.

A simple wave of my arm is enough to generate a breeze. The sweat evaporates off my skin, producing a cooling effect as energy is transferred, transforming the salty liquid water into a gas. That is the only comfort I'm likely to get from the heat today, but still I am awed by the process. In an hour or a week, those same beads of moisture, swept off my skin by the wind, will reappear as part of a cloud far over the horizon, seen by somebody else's eyes. What will that cloud tell them? Will they stop to marvel that a tiny portion of that airborne mist was once a part of me?

I wonder. How many neighbors and strangers, how many rivers and streams, how many plants and animals contributed to the clouds I see now? How many molecules of Caesar's dying breath occupy that cumulus cloud on the horizon? Just as the weather is made of water, so too are we.

A minuscule percentage of each cloud I see once sat as glaciers at the poles. Streams and lakes also contribute. But the vast majority of all water vapor comes from the oceans. And the cycle is endless.

"Some people are weatherwise, but most are otherwise," Ben Franklin is reputed to have joked. The other, more famous remark by Mark Twain is more apt. It bears repeating: "If you don't like the weather here, wait a minute." The sky changes. So does the climate. We cannot stop it, but we can strive to understand the hidden mechanisms that force the wind to dance and swirl.

Even now a southeast breeze tousles my hair. A thickening layer of stratus clouds advances overhead and indicates the arrival of a warm front. I glance up at the sky and prepare for rain.

.

FOR THE CONVENIENCE OF AMERICAN READERS, ALL TEM-
peratures, weights and distances in *Tying Down the Wind* are listed
using Fahrenheit degrees and the modified British Imperial system of
measurement. Most of the rest of the world now uses the metric system. Except in the United States and Burma, Fahrenheit degrees have
been replaced by Celsius, and units of distance like feet and miles appear almost as quaint as the obsolete *stadia* used by Eratosthenes in
200 B.C. In a way, that's a shame.

The decimal metric system makes calculations simple—in short,
it makes it easier to juggle numbers—and it is likely to overcome re-
sistance in the United States within a century. But the metric system
is not perfect.

The increasingly obsolete English system possesses a rustic
charm and some surprising advantages. Fahrenheit degrees are more
precise than Celsius degrees. And surely Robert Frost never would
have written, "Kilometers to go before I sleep." But let's compare the
two systems one unit at a time.

Distances and Weights

A FONDNESS for the number 12 permeates the old English system
of measurement. There are 12 inches to a foot, 12 units in a dozen, 12

dozen to a gross—even 12 shillings to the British unit of currency, the pound. To a mathematician with only ten fingers, the repetition of all those twelves can be maddening. (Just to make things even more confusing, there are 16 ounces in a pound; however, there were 12 ounces in the original Troy pound.)

The unit of distance called the foot, as its name suggests, was based on the length of a foot from heel to toe. The ancient Romans first divided it into increments of twelve. The Romans also invented the mile, defining it as the distance covered by 1,000 paces. The Latin word *mille* means "a thousand."

The problem, of course, was that no one's foot or pace was exactly the same. With slightly more precision (and also a little arrogance) King Henry I of England in the year 1100 defined a "yard" as the distance between the tip of his nose and the end of his outstretched thumb. More than a century later, Edward I decided that a yard should equal three feet. Eventually, one mile became equal to 5280 feet.

Confusing? Yes. The metric system, by contrast, clears things up and is more practical. But despite its flaws, the convoluted English system of weights and measurements has thoroughly permeated the English language and will take centuries to be removed. "Give them an inch and they'll take a mile," will probably survive as an adage long after anyone remembers what an inch or a mile represents. Translate that sentence to "Give them a centimeter and they'll take a kilometer," and it lacks the same appeal.

"Inchworms" crawl along the leaves. A heavy object "won't budge an inch." And if you've ever found an idea hard to fathom, consider that a "fathom" originated as a nautical unit of depth equal to six feet. Today, most people use the word only as a verb, meaning "to understand."

So even after it is fully supplanted by the metric system, the old English system is likely to linger in our everyday speech for many years to come.

Temperature

ALTHOUGH the first real thermometer was invented in 1654, it was nothing but an unmarked tube containing liquid that rose and fell as the temperature changed. No degrees or increments existed until 1701, when Isaac Newton suggested marking the tube "0" at the melting point of ice and (predictably enough, since Newton lived in England) 12 at body temperature.

In 1714, the German scientist Gabriel Fahrenheit replaced the existing water and alcohol thermometers with a mercury-based instrument. Not only did he improve the accuracy of thermometers, he expanded the range of the instrument. (A water-based thermometer obviously cannot measure temperatures below the freezing point of water, and alcohol will boil on a hot summer day.)

Fahrenheit set his "zero" at the lowest temperature he could create in his laboratory, which was equivalent to a fiercely cold winter night. At first he set body temperature equal to 12, as Newton had done. But Fahrenheit's thermometers were so sensitive that he decided to divide his scale into much finer increments. Keeping zero where it was, the freezing point of water on the Fahrenheit scale turned out to be 32 degrees, and the boiling point of water, 212. The difference between the two was a perfect 180 degrees, a number easy to work with mathematically (half a circle, for instance, is 180 degrees of arc), so Fahrenheit was pleased.

In 1742, a Swedish astronomer named Anders Celsius created a new scale for the mercury thermometer. Celsius set the boiling point of water equal to zero and the freezing point at 100. (A year later he reversed these numbers, so that the temperature went up instead of down as heat increased.) The interval between freezing and boiling was thus a convenient 100 degrees. He called his invention the Centigrade scale, derived from the Latin for "a hundred steps." Today the scale is named in honor of its inventor, and we speak of "degrees Celsius."

Though most of the world uses the Celsius scale, the Fahrenheit scale may be better suited to meteorology. For one thing, it is more precise simply because each degree represents a smaller interval.

More importantly, the range in temperature from zero to 100 degrees Fahrenheit almost perfectly demarcates the extremes found in the climates of the United States and Europe; the temperature seldom gets any hotter or colder. The convenience of a perfect 100-degree interval encompassing the temperatures in which most of us live seems a pity to lose. (The same range on the Celsius scale is a clumsier minus-18 to 38 degrees.)

However, the advantages of other aspects of the Celsius scale will win out in the end. (For instance, a Celsius degree is the same "size" as a degree Kelvin, making conversions and calculations much easier. Zero on the Kelvin scale equals absolute zero—the coldest temperature theoretically possible.) And so, in the future, a forecast of "ten degrees below zero" will not be as cold as it once was. I'm sure we'll get used to it.

TEMPERATURE CONVERSIONS

To convert Fahrenheit to Celsius, subtract 32, then multiply the result by 5/9ths (or 0.5555).

Example: 50 degrees F minus 32 = 18
          18 times 0.5555 = 10 degrees C

To convert Celsius to Fahrenheit, first multiply by 9/5ths (or 1.8), then add 32 to the result.

Example: 10 degrees C times 1.8 = 18
          18 + 32 = 50 degrees F

Comparisons of common temperatures:
    100 F = 38 C
     80 F = 27 C

40 F = 4 C
32 F = 0 C
−20 F = −28 C
−40 F = −40 C *(The two scales are equal at this point.)*
−60 F = −51 C

Extremes:

136 F = 58 C *(Hottest temperature on Earth, el-Aziziah, Libya, 1922.)*
134 F = 56 C *(Hottest temperature in United States, Death Valley, 1913.)*
−79.8 F = −62 C *(Coldest temperature in United States, Prospect Creek, Alaska, 1971.)*
−128.6 F = −89 C *(Coldest temperature on Earth, Vostok Station, Antarctica, 1983.)*

ENGLISH/METRIC CONVERSIONS

1 mile = 1.61 kilometers (1,609 meters)
1 yard = 91.44 centimeters (.9144 meter)
1 foot = 30.48 centimeters (.3048 meter)
1 inch = 2.54 centimeters (.0254 meter)

1 pound = 0.45 kilograms
1 ounce = 28.25 grams
1 gallon = 3.8 liters

METRIC/ENGLISH CONVERSIONS

1 kilometer = 0.62137 mile (3,281 feet)
1 meter = 3.281 feet (39.37 inches)
1 centimeter = 0.3937 inch
1 kilogram = 2.205 pounds
1 gram = 0.0353 ounce
1 liter = 0.2642 gallon (2.113 pints)

SPEED CONVERSIONS

1 mile per hour = 0.4470 meter per second
74 miles per hour = 33.08 meters per second
100 miles per hour = 44.70 meters per second

1 mile per hour = 1.609 kilometers per hour
74 miles per hour = 119.1 kilometers per hour
100 miles per hour = 160.9 kilometers per hour

1 mile per hour = 0.8690 knots
74 miles per hour = 64.3 knots
100 miles per hour = 86.90 knots

1 knot = 0.5144 meters per second
64.3 knots = 33.08 meters per second
100 knots = 51.44 meters per second
1 meter per second = 2.237 miles per hour
= 1.944 knots

1 kilometer per hour = 0.6214 miles per hour
= 0.5340 knots

ADIABATIC COOLING

The cooling of air due to expansion. If a parcel of air rises into the sky, it will increase in volume ("spread out") as the atmospheric pressure lowers. This causes the temperature to decrease as elevation increases. In "dry," unsaturated air, the temperature drops an average of 5.4 degrees F (3.5 degrees C) for every 1,000-foot rise in elevation. This is known as the dry adiabatic lapse rate.

ADIABATIC WARMING

The heating of air due to compression. If a parcel of air warms or cools adiabatically, it does so "internally," with no energy added or taken away. See *adiabatic cooling.*

ADVECTION

The transfer of moisture and heat via the horizontal movement of air.

AIR MASS

A large body of air distinguished by nearly uniform temperature and humidity at a given altitude.

ALBEDO

The percentage of sunlight (or any radiation) reflected off a specific surface. Fresh snow has an albedo of 80 to 90 percent; that is, snow reflects most of the incoming sunlight back into space. Thick clouds possess a similar albedo. Green grass, on the other hand, reflects only ten to 30 percent of sunlight; the rest is absorbed as heat. The albedo of ocean water or dark forest is even lower, usually ten percent or less.

ALTIMETER
A type of aneroid barometer used to measure altitude.

ANEMOMETER
An instrument used to measure wind speed. The familiar, rotating "three-cup" anemometers are sufficient for most situations, but too fragile for extreme events such as hurricane-force winds or tornadoes. One alternative is the Pitot-static anemometer.

ANTICYCLONE
A region of high atmospheric pressure, usually characterized by fair weather. Surface winds flow outward from the center in a clockwise direction in the northern hemisphere, and counterclockwise in the southern hemisphere.

APHELION
The point in a planet's orbit at which it is farthest from the sun. Planetary orbits are ellipses, not perfect circles. Earth reaches aphelion each year on or about July 4; the distance between the sun and Earth at this time is 94,506,240 miles (152,060,540 km). The average distance between the sun and Earth is 93 million miles (150 million km). See also *perihelion*.

AURORA
A glow of light in the night sky caused by the solar wind's electromagnetic interaction with Earth's upper atmosphere. Also called the northern lights (aurora borealis) and southern lights (aurora australis).

BACKING WIND

A wind that shifts in a counterclockwise direction (e.g., from northeast to northwest).

BAROGRAPH

A device used to measure and record atmospheric pressure.

BAROMETER

A device used to measure atmospheric pressure.

BERNOULLI EFFECT

The increase in wind speed as when a moving air mass is squeezed between the roof of the tropopause and an intruding landmass such as a mountain. It is based on a theorem published in 1738 by the Swiss physicist Daniel Bernoulli, who originally noticed the effect in a fluid flowing through a pipe. If you narrow the opening through which a fluid is flowing, the rate of flow will increase.

BLIZZARD

Technically, a blizzard is a period of snowfall and/or blowing snow that is accompanied by wind speeds greater than 35 mph (56 kph) and severely restricted visibility.

BUYS BALLOT'S LAW

A simple meteorological rule of thumb devised by Dutch scientist Buys Ballot in 1857. It demonstrates how atmospheric pressure determines wind direction. In the northern hemisphere, if you stand with your back to the wind, low pressure is on your left and high pressure on your right. The directions are reversed in the southern hemisphere.

CHLOROFLUOROCARBONS (CFCs)

Synthetic compounds used as coolant in air conditioners and some refrigerators, and for other industrial purposes. If released into the atmosphere, they eventually break down into chlorine, which damages the stratospheric ozone layer.

CIRROCUMULUS

High-altitude clouds primarily composed of ice crystals, with bases at about 20,000 feet above sea level. These clouds are characterized by their vertical development. They form in a wavy pattern of white patches that resemble fish scales. The expression "Beware a mackerel sky!" refers to cirrocumulus clouds. They often herald the approach of a warm front and stormy weather.

CLIMATE

The average weather patterns of a given region, including temperature and precipitation, over a long period of time.

COLD FRONT

The leading edge of a cool air mass pushing underneath and displacing a warmer air mass. A cold front often brings brief, heavy showers, thunderstorms, and strong winds.

CONDENSATION

The transformation of a substance from a gas to a liquid.

CONDENSATION NUCLEI

Microscopic impurities in the air, such as specks of sea salt, smoke, or dust, on which water vapor can condense, forming clouds or fog.

CONDUCTION

The transfer of heat within a substance or between substances by molecular activity. Metal conducts heat very well, which is why a silver spoon placed in a hot cup of coffee will quickly become hot. For other methods of transferring heat, see *convection* and *radiation*.

CONTRAIL

The condensation trail produced by the exhaust of an aircraft near the top of the troposphere. Contrails last longest and grow widest if the relative humidity is high at the aircraft's altitude. In extremely dry air, contrails are likely to be thin and short-lived if they form at all.

CONVECTION

The transfer of heat by the movement of a gas or liquid. In meteorology, the term usually refers to the vertical rise of warm parcels of air, best illustrated by towering cumulus clouds and hurricanes.

CONVERGENCE

The compression and distribution of wind resulting in a net inflow of air to a given region.

CORIOLIS EFFECT

A deflection in the motion of wind and ocean currents caused by Earth's rotation. Wind appears to curve to the right in the northern hemisphere and to the left in the southern hemisphere.

CORONA

A bright circle of light surrounding the sun or moon caused by the diffraction of light through thin clouds containing small water droplets. This phenomena is different from a halo. The word also refers to the outermost layer of the sun's atmosphere.

CUMULONIMBUS

Literally, a precipitating cumulus cloud; a thunderstorm.

CYCLONE

A low-pressure system with surface winds flowing inward toward the system's center (in a counterclockwise direction in the northern hemisphere or clockwise in the southern hemisphere). Cyclones are usually characterized by cloudy skies and stormy weather.

DEPRESSION

An area of low atmospheric pressure.

DEW POINT

The temperature at which water vapor will condense to form fog if a given parcel of air is cooled at a constant atmospheric pressure. The

closer the dew point is to the current temperature, the greater the relative humidity.

DIFFRACTION

The bending of light as it passes the edge of an object, such as a water droplet.

DIURNAL

Occurring or active during daylight rather than at night. Also defined as recurring on a daily cycle, once every 24 hours. Scientists prefer the latter definition.

DIVERGENCE

The distribution of wind in such a way as to cause a net outflow of air from a particular area.

DOLDRUMS

The equatorial zone between 10 degrees north and 10 degrees south latitude. The intensity of the sun heats air and causes it to rise in this region. Winds at ground level are often light and listless.

DOWNBURST

A strong blast of cool, descending air from a severe thunderstorm, often producing damaging winds at the surface.

DOWNDRAFT

A descending stream of relatively cool air originating high in a cumulonimbus cloud and causing sudden changes in wind speed and direction. Particularly violent downdrafts are also called microbursts.

EL NIÑO

The periodic warming of surface water from the western equatorial Pacific to the coast of Peru, associated with a slackening of the easterly trade winds. El Niño (Spanish for "the little boy") occurs every few years, usually beginning in December, and can effect weather patterns worldwide.

EQUINOX
One of the two times each year when the sun crosses the equator and day and night are approximately equal at 12 hours apiece. The vernal equinox occurs on or around March 21 and heralds the start of spring on most calendars (meteorological spring runs from March 1 through the end of May). The autumnal equinox occurs on or around September 22.

EVAPORATION
The transformation of a substance from a liquid to a gas. Liquid water evaporates more quickly in warm air than in cold air.

EXOSPHERE
The outermost layer of Earth's atmosphere, beginning at an altitude of about 350 miles (560 km), higher than the orbit of the space shuttle. The exosphere is virtually indistinguishable from the vacuum of space. Hydrogen and helium, the lightest of all molecules, are the most common substances in the exosphere.

EYE WALL
The vertical wall of clouds and strong winds surrounding the calm central eye of a hurricane or tropical storm.

FERREL CELL
One of the two regions of prevailing westerly airflow at middle latitudes, between 30 and 60 degrees north and between 30 and 60 degrees south.

FRONT
A boundary between cold and warm air masses.

FROST

Ice crystals that form on an exposed surface, such as a windowpane or grass stem, by the sublimation of water vapor in cold air.

FUNNEL CLOUD

A rotating column of air and fog descending from the base of a cumulonimbus cloud toward the ground. If it touches the ground, it is classified as a tornado.

GLAZE

A hard, slick coating of ice formed by freezing rain or freezing drizzle.

GRAUPEL

Tiny white ice particles coated by thin layers of supercooled water that have frozen around the nuclei. Also called snow pellets or soft hail. This form of precipitation bounces when it strikes a hard surface.

GREEN FLASH

A brief greenish wink of light seen on the horizon at the moment of sunrise or sunset. It is caused by the atmosphere's refraction of green light, and can be seen best from a mountaintop on a clear, haze-free day, or from the ocean.

GREENHOUSE EFFECT

The process by which carbon dioxide, water vapor molecules, and methane in the atmosphere absorb shortwave solar radiation that has been reflected off Earth's surface; these molecules subsequently emit energy, so that the ground receives heat from both the sun and the atmosphere. Without the greenhouse effect, the planet would be locked in a perpetual ice age. Global warming is the term for a greenhouse effect taken to an uncomfortable extreme.

GULF STREAM

A warm ocean current flowing from the Gulf of Mexico to the coast of Europe. It helps to moderate temperatures in the North Atlantic and the climate of western Europe.

HADLEY CELL

One of the three bands of air circulation patterns in each hemisphere, powered by temperature and pressure differences due to the uneven heating of Earth by the sun. Hadley cells exist north and south of the equator at tropical and subtropical latitudes.

HAIL

Precipitating balls or clumps of ice produced by updrafts within a cumulonimbus cloud.

HALO

A ring around the sun or moon caused by the refraction of light through ice-crystal clouds. A halo is often a sign of impending stormy weather, hence the saying "Ring around the sun or moon, rain will come soon." The clouds that produce the arc or halo are often cirrostratus clouds on the leading edge of a warm front or a hurricane.

HAZE

Tiny impurities in the air, including pollen, sea salt, dust, and pollutants, that reduce visibility.

HIGH

A region of cooler, sinking air, characterized by fair skies, no precipitation, and high atmospheric pressure. See *anticyclone*.

HORSE LATITUDES

The becalmed ring of high pressure circling Earth between 30 and 35 degrees north latitude. The name originated when ships sailing between Europe and the Americas encountered calm winds for extended periods of time and ran low on supplies. To save food and water, the crews threw horses overboard.

HURRICANE

A tropical cyclone with sustained wind speeds of at least 74 mph (119 kph). Hurricanes are created by convection (rising, unstable air) over warm water. Unlike extratropical low-pressure systems, which are created by the clash of warm and cold air masses, hurricanes usually do not have fronts.

INVERSION

A situation in the troposphere in which temperature rises instead of falls as altitude increases, so that warmer air overlies a layer of colder air at the surface. Normally the reverse is true.

ION

An atom or molecule that possesses a positive or negative electrical charge, due to either the loss or gain of one or more electrons.

IONOSPHERE

A region of ionized gases in the atmosphere between about 30 and 250 miles above Earth's surface. The ionosphere overlaps the mesosphere and the thermosphere, and is the level of the atmosphere in which aurora occur.

ISOBAR

A line on a weather map drawn through points of equal atmospheric pressure. The greater the difference in pressure between one location and another, the closer the lines to one another.

ISOTHERM

A line on a weather map drawn through points with equal temperatures.

JET STREAM

A river of high-speed wind near the top of the troposphere. Wind speeds can exceed 200 mph in the center of the jet stream, but are slower along the edges. The jet stream is an important factor in the formation of storms and the directions in which they move.

KATABATIC WIND

A downslope flow of cold, dense air being pulled by gravity.

LA NIÑA

The periodic cooling of surface water from the western equatorial Pacific to the coast of Peru. (Spanish for "the little girl.")

LATENT HEAT

Energy absorbed or released when a substance such as water changes phase, e.g., from liquid to gas. The temperature is often noticeably cooler immediately after a rain shower because of the evaporation of raindrops; the water "uses up" heat and energy to change from a lower phase (liquid) to a higher one (gaseous water vapor).

LENTICULAR CLOUD

A lens-shaped cloud formed by the wave-like flow of wind. Often seen in mountainous regions.

LOW

A region of rising air characterized by cloudy skies, precipitation, and low atmospheric pressure. See *cyclone.*

MESOCYCLONE

A violently rotating spiral of air in a large thunderstorm that may produce a tornado. Mesocyclones are usually about ten miles in diameter.

MESOSPHERE

The coldest layer of Earth's atmosphere, approximately 30 to 50 miles above the surface.

MICROBURST

A strong, violent downdraft from a thunderstorm.

MONSOON

A seasonal wind blowing from land to ocean in winter, ocean to land in summer. The word is derived from the Arabic word for summer, *mausim*. Summer monsoons flowing off the Indian Ocean bring torrential rains to the Indian subcontinent.

OCCLUDED FRONT

A boundary between cool, cold, and warm air masses, sometimes formed when a cold front overtakes a warm front, forcing warm air aloft and producing an inversion.

OROGRAPHIC LIFTING

The forced uplift of air over a mountain or ridge. As the air rises, it cools and may reach its dew point, forming a cap cloud.

OZONE

An oxygen molecule consisting of three bonded oxygen atoms. (The oxygen breathed by animals has only two bonded atoms.) At ground level, ozone is a pollutant. High in the atmosphere, however, it absorbs harmful ultraviolet radiation and provides a necessary protective shield for life on Earth. The ozone layer is concentrated in the stratosphere. Depletion of this protective shield due to CFCs is thought to contribute to an increase in skin cancer and to other environmental hazards.

PARHELIA

See *sundogs*.

PERIHELION

The point in a planet's elliptical orbit at which it is closest to the sun. Earth reaches perihelion each year on or about January 2; the distance between the sun and Earth at this time is 91,406,076 miles

(147,072,376 km). The average distance between the sun and Earth is 93 million miles (150 million km). See also *aphelion*.

PHOTOSYNTHESIS

The process by which chlorophyll in green plants converts sunlight into energy. Oxygen is a by-product of this process.

PRESSURE GRADIENT FORCE

The force that results from the tendency of air to flow from an area of high pressure to one of low pressure.

RAINBOW

A colorful arc caused by the reflection and refraction of sunlight through falling raindrops. In order to see a rainbow, you must stand with the sun at your back and face a distant area of the sky in which rain is still falling.

REFRACTION

The bending of light as it passes from one transparent medium to another, e.g., from air to water. The two mediums must have different densities.

RELATIVE HUMIDITY

The ratio of the amount of water vapor in the air to the amount necessary to saturate the air at a constant temperature. At 100 percent relative humidity, air is saturated and fog or clouds will appear. The lower the percentage, the drier the air. Warm air generally contains more water vapor than does cold air per given area. 100 percent relative humidity at 80 degrees Fahrenheit is "wetter" than 100 percent humidity at zero degrees.

RETROGRADING

The movement of a weather system from east to west at mid-latitudes, instead of the more usual west-to-east direction.

RIDGE

An elongated area of high atmospheric pressure produced by a northward-flowing wave in the jet stream.

RIME

A feathery ice formed by supercooled cloud droplets that freeze on contact with solid objects. Rime grows into the wind and contains many air bubbles; it is softer than glaze.

SHORT WAVE

A bend or ripple in the jet stream, or other high-altitude wind, that may trigger a clash of cold and warm air masses, thus creating a storm.

SOLSTICE

The times each year when the sun is farthest north or south of the equator, producing the longest and shortest days of the year. In the northern Hemisphere, the summer solstice occurs on or about June 21. The winter solstice occurs on or about December 22.

SQUALL LINE

A line of cumulonimbus clouds and unstable air as much as 250 miles across, often bringing strong winds, heavy showers, and hail. Squall lines form in advance of cold fronts.

STATIONARY FRONT

A boundary between two air masses when neither the cold nor warm air mass is making any headway; the front doesn't move, hence its name. Cloud cover is often widespread on both sides of a stationary front.

SUBLIMATION

The process by which a solid changes directly to a gas, or vice versa, without passing through the liquid phase. Frost that forms on grass or car windows is often a result of the sublimation of water vapor.

SUNDOGS

An optical phenomenon caused by the refraction of light through ice-crystal clouds. It is the appearance in the sky of two bright spots to the right and left of the sun. Also called mock suns or parhelia.

SUPERCELL

A large, rotating thunderstorm, or mesocyclone, frequently associated with tornadoes.

SYNOPTIC CHART

A map listing the various weather conditions at different locations at a specific time and date.

THERMOSPHERE

A layer of Earth's atmosphere, more than 50 miles high, in which temperature increases with height. Air pressure is extremely low in the thermosphere.

TORNADO

A violently spinning column of air reaching from the base of a cumulonimbus cloud to the ground. Wind speeds in a tornado can range from 40 mph to an estimated maximum of over 300 mph. (No anemometer has ever survived a strong tornado's fury to provide an official measurement.) Tornadoes usually move across the ground from southwest to northeast at an average speed of 35 mph. Their vortices are generally about 100 yards in diameter, though some tornadoes have been more than a mile wide.

TRADE WINDS

The prevailing east-to-west-flowing winds that exist north and south of the equator in tropical latitudes.

TRANSPIRATION

The process by which plants release water vapor into the atmosphere.

TROPOPAUSE

The atmospheric boundary between the troposphere and the mesosphere at an altitude of about five to ten miles. Above this "ceiling," the temperature of the atmosphere is steady or increasing. That prevents warm air currents from rising any higher (see *convection*) and limits the heights of most clouds to the lowest layer of the atmosphere.

TROPOSPHERE

The lowest, densest layer of Earth's atmosphere. It extends from the surface up to a maximum altitude of ten miles. Almost all clouds, water vapor, and weather are contained within the troposphere.

TROUGH

An elongated area of low atmospheric pressure, produced by a southward-flowing wave in the jet stream.

VEERING WIND

A wind that shifts clockwise in direction (e.g., from southwest to northwest). Often this occurs during a frontal passage.

VIRGA

Rain or snow falling from a cloud that evaporates before reaching the ground.

WARM FRONT

The leading edge of a warm air mass displacing or overrunning a cooler air mass. A warm front brings thickening and lowering clouds. Precipitation is usually steady, rather than showery and intermittent.

WATER VAPOR

Water in gaseous form in the atmosphere.

WIND SHEAR

A sudden change in wind speed and direction. See also *microburst*.